AF335356

Don Giovanni
CAPTURED

Performance, Media, Myth

RICHARD WILL

The University of Chicago Press   Chicago & London

The University of Chicago Press, Chicago 60637
The University of Chicago Press, Ltd., London
© 2022 by The University of Chicago
All rights reserved. No part of this book may be used or reproduced in any
manner whatsoever without written permission, except in the case of brief
quotations in critical articles and reviews. For more information, contact the
University of Chicago Press, 1427 E. 60th St., Chicago, IL 60637.
Published 2022
Printed in the United States of America

31 30 29 28 27 26 25 24 23 22   1 2 3 4 5

ISBN-13: 978-0-226-81541-1 (cloth)
ISBN-13: 978-0-226-81542-8 (e-book)
DOI: https://doi.org/10.7208/chicago/9780226815428.001.0001

This book has been supported by the Joseph Kerman Fund and the General
Fund of the American Musicological Society, funded in part by the National
Endowment for the Humanities and the Andrew W. Mellon Foundation.

Library of Congress Cataloging-in-Publication Data

Names: Will, Richard James, author.
Title: "Don Giovanni" captured : performance, media, myth / Richard Will.
Other titles: Opera lab.
Description: Chicago ; London : The University of Chicago Press, 2022. |
Series: Opera lab: explorations in history, technology, and performance |
Includes bibliographical references and index.
Identifiers: LCCN 2021032886 | ISBN 9780226815411 (cloth) | ISBN 9780226815428 (e-book)
Subjects: LCSH: Mozart, Wolfgang Amadeus, 1756–1791. Don Giovanni. | Mozart,
Wolfgang Amadeus, 1756–1791—Performances. | Mozart, Wolfgang Amadeus,
1756–1791—Discography. | Operas—Production and direction—History. | Operas—
Discography. | Operas—Film catalogs. | Operas—Film adaptations.
Classification: LCC ML410.M9 W6218 2021 | DDC 782.1—dc23
LC record available at https://lccn.loc.gov/2021032886

♾ This paper meets the requirements of ANSI/NISO Z39.48-1992 (Permanence of Paper).

Contents

✳ Note to Readers

The companion website for this book (http://press.uchicago.edu/sites/ will) includes color images of figures 1.2 and 5.1 as well as audio playlists for the introduction and chapters 1-7, which are referenced by a symbol in the text (♫). The video clips discussed in chapters 8–10 are available on published video discs and through streaming services. When referring to complete audio and video recordings, I use title, conductor (audio) or stage director (video), and date of recording (e.g., *Don Giovanni*, cond. Busch, 1936). Details for all recordings, including dates of release, are in the discography and videography.

Unless otherwise noted, translations are my own. Quotations of the Italian libretto follow the critical edition by Giovanna Gronda, *Il Don Giovanni: Dramma giocoso in due atti* (Turin: Einaudi, 1996).

✳ Tables

*

Figures

*

Introduction

A few years ago I started watching and listening to as many recordings of *Don Giovanni* as I could find. I had wanted to do so for a long time, ever since the opera roiled my first efforts at teaching Western music history in the early 1990s. Like a dutiful historian I focused on origins, showing how Mozart and his librettist, Lorenzo da Ponte, followed comic-opera convention by depicting romantic intrigue amongst aristocrats and their subjects, and also stretched convention by choosing an intrigue involving rape, murder, and divine retribution. I framed their subject in terms of eighteenth-century sexual politics, and I worked hard to illustrate how they brought the story to life, especially Mozart through his celebrated representation of action and feelings. The classes were up to date, thorough, and effective, or so I thought until I introduced another standby of music history teaching: audio and video illustration.

I knew perfectly well that audio and video recordings preserve differing interpretations, and so did my students. But in the early 1990s *Don Giovanni* existed in such disparate versions that "interpretation" seemed like the wrong word. In some, men in tights romped around Renaissance villas; in others, guys in suits (or jeans) skulked through modern cities. In some, the libertine protagonist was a lovable rake; in others, he was a vicious predator. In some, big voices sang long lines over vast orchestras; in others, not-so-big voices turned crisp phrases over little bands of antique instruments. Several changes were going on at once. *Don Giovanni* had been getting more serious since the 1970s, as the feminist struggle against rape and the objectification of women called into question the long-standing treatment of its story as comic or heroic. At the same time Mozart's operas got caught up in artistic conflicts, on the one hand between conventional stagings and revisionist practices such as modernized settings, and on the other between mainstream classical music performance and the movement best known these days as "historically informed performance." The tensions were exposed by a flood of recordings, prompted by the 1991 Mozart bicentennial as well as by the spread of home video and CD players. Between 1985 and 1995 *Don Giovanni* appeared on a dozen newly recorded CDs, at least a dozen CD reissues of vintage LPs, and several VHS tapes of both recent and historical productions.[1] All at once, students, teachers, and anyone else with the right audiovisual gear could experience several decades' worth of divergent interpretation.

It was exhilarating—and paralyzing. The *Don Giovanni* of my classes had a reassuring stability rooted in the score, the libretto, and the social mores of eighteenth-century Europe. How could I square this with its instability in performance? Should I double down on the origin story and choose my illustrations for their *Werktreue*, a German coinage meaning "fidelity to the work"? Teachers of classical music do this a lot, recommending authentic manuscripts and creators' intentions as the best guides to interpretation. Too much of an historian to disagree entirely, I nonetheless had misgivings: I knew that original sources can be very slippery, and that Mozart and Da Ponte had themselves left more than one legitimate version of *Don Giovanni*.

I also worried that my students, some of them hopeful of operatic careers, needed more information to make sense of this particular work than eighteenth-century sources could provide. The new stagings and musical styles of the 1980s and 1990s were doing more than spruce up a warhorse: they were reshaping and sometimes rejecting a mythology. For two centuries, performers and commentators had taken *Don Giovanni* to express verities about sex and eros. Some voiced frankly sexist notions—most notoriously that the opera's title character embodies demonic or biological forces that render his actions beyond moral judgment. Others downplayed the violence of those actions, calling the same character a seducer but not a rapist or a murderer, an omission encountered in textbooks, among other places.[2] Teaching only the "original" opera would leave students ignorant of this thorny legacy, and of a counterreaction that was beginning to knock the libertine off his pedestal. By the time I started teaching, *Don Giovanni* was a creation of the nineteenth and twentieth centuries as much as the eighteenth; a Romantic, modernist, and postmodernist opera as much as a classical one.

So I began to wonder about upending my approach. Rather than just one *Don Giovanni*, what if I taught several, including but not limited to the original? Why not trace the opera's progress as a subject in itself, turning my audio and video supplements into the main attraction? This could highlight the social and political stakes of reenacting such a story, and the role of performers and commentators in determining its meaning. The idea was appealing, but practical concerns gave me pause. It was one thing to talk about an opera, particularly one with a long and well-documented trail of source studies and commentaries.[3] It was quite another to tackle performance, even just recorded performance. I could devote the whole term to *Don Giovanni* and still have time for only a fraction of the available recordings, and in truth I was scarcely qualified to pick and choose. I knew enough releases to recognize their variety, but they represented a tiny fraction of an archive encompassing hundreds of performances stretching back more than ninety years. And there was a further challenge: by comparison with Mozart scholarship, the literature on recordings was wickedly decentralized, full of insight but scattered across newspaper and magazine reviews, record guides, collector catalogues, and performer memoirs. If I really wanted to teach "*Don Giovanni* through the ages," a lot of excavating lay ahead.

Vantage Points

Which brings us to this book. Three decades on, the prospects for studying an opera's performance history have improved. The number of readily accessible recordings has continued to grow: since 1995, commercial releases of *Don Giovanni* have included another twenty-five videos, several new and dozens of reissued complete audio recordings, and a wide array of reissued excerpts from the era of wax cylinders and 78 rpm discs. Especially with the rise of the Internet, numerous unreleased versions have circulated as well, notably radio and television broadcasts from the 1960s, '70s, and '80s. The digital revolution has made navigating recordings easier, a matter of clicking around in files rather than rewinding tapes or swapping discs, and new software has brought to light previously unmeasurable nuances of tempo, vibrato, vocal color, and more. Though late in coming, a burgeoning field of performance research has also begun to establish some context, charting the sea changes in singing, playing, and staging that have shaped every opera in the decades since 1900.

These resources show that the variety of interpretation I encountered in the early 1990s was hardly unusual. Performers have never agreed on how *Don Giovanni* should go, and they still don't.[4] At the dawn of the recording era, around 1900, some of the most exuberant performances ever captured exist side-by-side with some of the most restrained, reflecting generational shifts as well as Romantic and Victorian differences over the opera's subject matter. Fifty years later, by midcentury, restraint prevails, but a fresh contrast emerges between gravity and urgency, between the opera's tragedy and supernaturalism on the one hand, and its comedy and psychological realism on the other. Fifty years after that, at the turn of the twenty-first century, realism prevails along with a new skepticism of libertines, but in settings where the characters may look less like the authors of their own destinies than the victims of larger social or psychological ills. These are only the most obvious disparities, supplemented by 120 years' worth of subtle differences in musical performance, acting, stage design, costuming, and more.

Complicating matters further, even the most epochal changes do not always follow the kind of historical schedule that makes for tidy book narratives or course syllabi. Certain trends develop clearly over time: for example, the early twentieth-century decline of Romantic practices such as tempo rubato and improvised ornamentation, or the late twentieth-century rise of historically informed performance. But the variety within trends is so pronounced, and so unevenly distributed, that performance history has to be viewed from vantage points other than just chronological evolution. In the case of *Don Giovanni*, a useful starting point is characterization. The underlying legend is straightforward: "Don Juan is a libertine, so Heaven punishes him," in Wye Jamison Allanbrook's blunt précis.[5] Developed into a full-blown opera, the tale draws much of its power from the nature of the people involved, and this is as true for audio and video recordings as it is for stage performances. Sometimes the characters come across as social types, libertines and ingenues and *donne abbandonate*; other times they suggest the individuality and inner complexity of modern, Freudian subjects. Sometimes their personalities develop; other times they remain woodenly fixed. Sometimes their actions invite sympathy; other times they trigger revulsion. This last distinction determines the moral compass of many performances, especially by way of the title character. When played as a sexual predator, he is objectionable by any standard; but as an elegant rake, he may seem attractive despite his wrongdoings—and as a victim of social or psychological circumstance, he may even seem innocent. The other characters vary equally in their rectitude, whether it is the women acting more resistant or complicit, the servant Leporello more enthusiastic or reluctant, the noble lover Don Ottavio more self-sacrificing or just plain selfish.

A complementary, equally illuminating vantage point concerns the passage of time. Rooted in religious drama, the story is partly that of a race against the clock, an effort to squeeze in as many adventures as possible before the inevitable reckoning. When the latter comes, it is described in terms of the clock running out: "Ah tempo più non v'è" (Ah! there is no more time left to you),[6] sings the avenging statue to Don Giovanni, echoing his predecessor in the earliest literary realization of the legend, Tirso de Molina's *El burlador de Sevilla* of 1630.[7] As experienced in the opera, time is both the metaphorical power that brings down the libertine and the tangible, musical-dramatic quality that differentiates every scene both within performances and between one performance and the next. That same confrontation between statue and libertine, one of the most famous scenes in all of opera, lasts twice as long in some renditions as in others, and its moment-to-moment progress ranges from perfectly steady to disorientingly uneven. Whether the point of comparison is raw speed, degree of flexibility, or the trajectory from one section to the next, the flow of musical-dramatic time projects different visions: alternative images of fate itself.

No less valuable is the vantage point of recording itself, both the process and the institution. Strictly speaking, recordings are not part of performance, unless one counts productions that incorporate onstage video screenings or projections. Yet they are critical to how we perceive operas and their history. Media companies, together with opera houses and broadcasters, determine which performances get disseminated, and the industry favors the most prominent houses, festivals, and artists, mainly European and North American. Given their wide circulation and role in spawning ingrained assumptions about the opera, such "mainstream" releases make a logical focus for a book like this. Nonetheless, they enshrine a hierarchy dominated by well-capitalized Western firms and venues, leaving out of account the achievements of innumerable less-privileged houses and artists around the globe.

Recordings also determine what we see and hear of a performance. Audio recordings include no visuals; videos show only what camera directors choose; wax cylinders and 78 rpm discs break operas apart into their most salable excerpts. More subtly, successive generations of technology alter the nature of what we experience, almost as much as successive generations of performers do. The music industry long pursued its own ideal of "high fidelity," aiming in theory for the most perfect reproduction of live sound: German audiophiles call it *Klangtreue*, a suggestive parallel to *Werktreue*. In practice, the quest has produced competing ideals, from acoustic grittiness to digital noiselessness to streaming-ready compression, each with its own enthusiasts (and detractors) and its own way of representing voices and instruments. In film and video, successive shifts from black-and-white to color, and from analogue to digital, have left comparably divergent legacies. Time, character, and everything else that makes up an opera are inseparable from the media envelope in which they arrive.

Given these constraints, readers may wonder why I have chosen to study *Don Giovanni* "captured" rather than "live." The versions I have witnessed onstage, each with its own take on the opera and its myth, have undoubtedly influenced the discussions below. Still, my reasons for studying recordings go beyond the obvious challenge of writing about live performances that readers cannot experience for themselves. As studies of *Don Giovanni*'s eighteenth- and nineteenth-century history demonstrate, the documents of performance—mostly printed, in that era—show how operatic myths are made.[8] Recordings are no more complete than those earlier sources, yet they carry the story forward, affording glimpses not just of epochal changes in characterization or social politics, but also of day-to-day decisions through which singer-actors, instrumentalists, conductors, designers, and directors bring those changes about. For all their limitations, printed and recorded archives alike illustrate how messy the business of interpretation can be, how creative in its own right, and how consequential for a work's reputation.

Infidelities

That *Don Giovanni* has changed so much owes something to theatrical and musical performance generally, and something to its own texts. Operatic sources provide less information than the cachet of *Werktreue* would suggest. For starters, eighteenth-century scores and libretti say relatively little about staging. The culminating scene for Don Giovanni and the statue carries a single brief description of the set: "Sala . . . una mensa preparata per mangiare" (Hall . . . a table set for eating). Even if the designer opts for Renaissance Spain—the original rather than an updated locale—everything about the hall and table is left to the imagination: shape, size, arrangement, color, lighting, decoration. Stage directions are equally laconic: after the statue utters his fateful condemnation, the libretto calls for "foco da diverse parti tremuoto etc. . . . il foco cresce D. Gio. si sprofonda" (fire from several places, quaking, etc. . . . the fire increases, D[on] Gio[vanni] sinks down). "Etc." leaves considerable leeway even for performers determined to be faithful to the original, and it has accommodated plenty of invention over the years.

The musical notation offers more guidance but still leaves basic questions unanswered. Mozart lived before the invention of the metronome, and his tempo indications are necessarily approximate and relative. He marks lots of articulations, but the precise shapes of notes and phrases—length, accent, shades of volume, all the things that make a performance distinctive—are beyond the reach of notation. So is tone color: the score can specify instruments and vocal ranges but not how they will actually sound, and as any opera fan knows, the differences can be vast. Add to this the evidence that eighteenth-century musicians habitually ornamented what they saw on the page, and one is tempted to agree with the Shakespeare performance scholar William B. Worthen, who writes of the plays: "Most of what happens even in a conventional performance has no specification in the text at all."[9]

With *Don Giovanni* the ambiguity is compounded by its survival in multiple versions. Following its 1787 premiere in Prague, Mozart and Da Ponte wrote several new numbers for a run of performances in Vienna the following year. They include an aria for Don Ottavio, "Dalla sua pace"; an accompanied recitative and aria for Donna Elvira, "In quali eccessi, o Numi . . . Mì tradi quell'alma ingrata"; and a duet for Leporello and Zerlina, "Per queste tue manine," with surrounding recitatives. Studies of the sources (a rich tangle of the composer's autograph score, other scores and orchestral parts, and published libretti) show that the authors contemplated several ways of sequencing the new numbers with the old, and of cutting or reordering the Prague music to accommodate them.[10] Yet Mozart never identified a given sequence as definitive, and performances of the later eighteenth and nineteenth centuries adopted a wide variety of solutions.

None of this is unusual for the era, when composers and their collaborators expected to adapt their works for new performers and audiences. As Ian Woodfield writes, "An eighteenth-century opera did not consist of a fixed text; rather it was a fluid, constantly evolving enterprise, subject to a significant (and growing) element of collective responsibility."[11] Somewhat less typical is the extent to which differing versions of *Don Giovanni* have continued to compete for authority in more recent times. The most familiar retains the music and ordering from 1787 and adds the two new arias, Don Ottavio's in act 1 and Donna Elvira's in act 2. In the first half of the twentieth century especially, performers of this version often trimmed or omitted the ensemble scene following Don Giovanni's death as well, another revision that originates in 1788. The resulting composite underpins countless productions down to the present day, as well as the majority of complete audio and video recordings (the recordings usually with final scene intact). At the same time, two alternatives have gained more and more currency, satisfying a desire to disentangle the opera's stages of development and in some cases to mitigate perceived weaknesses in the Vienna revisions.[12] One consists of the music from 1787 without additions; the other, of what may have been the final sequence decided upon in Vienna, incorporating all three of the new numbers while cutting two original arias along with the concluding ensemble.[13] Their separation was no doubt encouraged by the *Neue Mozart Ausgabe* (the new complete edition, prepared by specialists and intended to be authoritative), whose *Don Giovanni* volumes of 1968 put the Prague music first and relegated the Viennese music to an appendix. Though not unheard of in earlier years, like-minded performances and recordings became increasingly common thereafter, especially as the historically informed performance movement sparked fresh interest in presenting works in their "original" form. Today what are commonly referred to as the Prague and Vienna versions are viable alternatives for productions of all stripes, with or without historicizing intentions.

Whatever the version, *Don Giovanni* has been further destabilized by its subject matter and plot. Mozart's contemporaries complained that he had wasted his music on an immoral, silly topic: "A pity it does not eat as well," wrote one wag about the statue, "for only then would the fun be complete."[14] The opera may have gone on to legendary status, but its history is littered with efforts to motivate, ameliorate, or flat-out eliminate the most offensive aspects of the story. Its sheer confusedness has inspired additional revisions. If the legend is bare, Da Ponte's adaptation abounds in puzzles, among them bewildering scene changes and the remarkable apparition of a completed stone monument to the Commendatore just hours after his murder. Neither incoherence nor immorality is unique to *Don Giovanni*, but their severity helps to explain why the critic David Littlejohn, writing in 1981, would call it "the impossible opera."[15]

To make the impossible real, performers cannot rely on sources alone. They have to fill in the gaps, take a stand on the subject matter, and sort out the dramatic inconsistencies, all while striving to satisfy the musical and theatrical expectations of their own day. The effects are felt everywhere, beginning with staging, where sources say the least and fashions change the most. *Don Giovanni* has been set in numerous historical periods, from the Renaissance to the twenty-first century, with scenery ranging from ornate to abstract, and stage action from static to frenetic. Less often remarked but equally significant, the opera's very words have changed, at least as they are experienced onstage or in recordings. Until 1950, performances outside of Italy were frequently given in translation, and while the practice has waned, it has never disappeared entirely. Apart from their impact on sound, so-called singing translations can introduce startling changes: the most widely used German text of the nineteenth and early twentieth centuries, by Friedrich Rochlitz, expunged nearly all of Da Ponte's many sexual innuendoes and gave the protagonist a surprising philosophical bent.[16] Though not part of the sounding text, the running translations now used as sub- or supertitles in theaters and video recordings can also have an effect. Many videos signal their own moral compass through the subtitling of a scene, early on, in which Leporello reproaches Don Giovanni for killing the Commendatore:

1787:   Bravo: Due imprese leggiadre! sforzar la figlia ed ammazzar il padre.

1995:   Two clever tricks: rape the girl and murder her father.

2002:   That's a fine double slaughter. Murder the father when you've ravished the daughter.

2010:   Well done. Two wonderful deeds: ravish the girl and kill the father.[17]

In the choice of "deeds" versus "tricks," "ravish" versus "rape," or "kill" versus "murder," to say nothing of the sarcastic rhyming of "daughter" and "slaughter," each version registers a slightly different degree of outrage.

As for the story, *Don Giovanni* has seen its share of wholesale adaptations, from an 1834 Parisian version in which the title character dreams of his impending doom in a ballet sequence (and Donna Anna, who has killed herself for having loved him, is buried to the strains of Mozart's Requiem),[18] to Jesusa Rodríguez's 1987 *Donna Giovanni*, a milestone of the bicentennial years in which a mostly female cast takes turns playing the protagonist.[19] More commonly, performers modify the original text—again, in whichever version—with cuts and tweaks, sometimes small but always significant. Omitting or trimming recitatives reduces the opportunity for characterization and sometimes scrambles the plot still further (see chap. 6). Cutting the final scene, as may be done even in the composite version, prevents the surviving characters from reacting to Don Giovanni's death, tilting the balance from comedy toward tragedy and from realism toward the supernatural. Having the others physically kill Don Giovanni, as in a few recent productions, ejects the supernatural altogether and turns the statue into a figment of his imagination. It also deepens the moral confusion, empowering his victims only to ensnare them in murderous revenge.

Musical performance varies no less. As already noted, tempos have waxed and waned dramatically, and the same can be said for everything from dynamics to tone color to accentuation. Ornamentation makes for an especially instructive example, inasmuch as performers have reversed course since 1900—twice. Singers trained in the nineteenth century routinely spiced up Mozart's vocal lines, often by adding bravura high notes to final cadences. Numerous early recordings bear witness, as do the objections of critics such as Bernard Shaw, who wrote of an 1891 *Don Giovanni*: "I am sorry to add that alterations of Mozart's text were the order of the evening, every one of the singers lacking Mozart's exquisite sense of form and artistic dignity."[20] Shaw was no paragon of fidelity: he recommends elsewhere in the same review that performers shorten the recitatives in order to make room for more of his favorite arias. Still, enough people shared his distaste for ornamentation that by 1930 it had all but disappeared from Mozart, not only added high notes but also trills and other decorations that could have come straight out of eighteenth-century manuals for music instruction. For fifty years the composer's melodies grew plainer, his cadences more decorous. Then the manuals were rediscovered, ornamentation returned, and critics changed their tune. A century after Shaw, the *Los Angeles Times*'s Herbert Glass would complain that the performers in a historically informed *Don Giovanni* did not spice things up enough: Della Jones, the Donna Elvira, was "the only singer in the cast of this 'authentic' presentation consistently interested in applying ornaments."[21] *Werktreue* falters because the objects of fidelity, the very sources and practices that performers are supposed to be faithful to, are themselves moving targets.

Souvenirs

It is scholars who have tried hardest to pin them down, whether by producing meticulously documented scores, or by researching practices such as improvised ornamentation. By the 1980s they seemed to be succeeding, as the *Neue Mozart Ausgabe* and other post–World War II scholarly editions neared completion and some historically informed performers professed an allegiance to document-based interpretation. The musical consequences were profound, leading to the extreme contrasts I encountered in the early 1990s, and eventually to new standards for the performance of eighteenth- and early nineteenth-century music altogether. The results for scholarship were equally significant, but not in the way one might have expected. Fealty to documents provoked a counterreaction, most significantly a series of articles and reviews by Richard Taruskin.[22] While enthusiastic in some cases about the musical results of historically informed practice, Taruskin argued that its intellectual claims did not hold up. As numerous examples of insufficient or misinterpreted evidence showed, no amount of research would allow the music of Mozart, Bach, or Beethoven to be played as it was in the composers' lifetimes. On the contrary, the "historical" style derived from more recent models: principally, in Taruskin's view, a performance tradition inaugurated by Stravinsky and other early twentieth-century modernists, the chief characteristics of which are steady tempos, crisp articulation, transparent textures, and—not least—strict adherence to the composer's text. By adopting similar values for older repertoires, performers were not so much restoring the originals as reinventing them to suit contemporary tastes. Taruskin's critique had its critics, but it opened new vistas by suggesting that interpretive trends deserved to be historicized in their own right, not simply praised or condemned according to their fidelity.[23] Equally significant, Taruskin used audio recordings to make his case, notably in CD reviews, where he juxtaposed period-instrument renderings with earlier versions of modernism as well as various antimodernisms. We shall see that his favorite example of the latter, the conductor Wilhelm Furtwängler, has a special role in the history of *Don Giovanni* performance.

As Taruskin scrutinized historically informed efforts, others took new interest in classical recordings generally, raising more questions about *Werktreue*. Traditionally, scholars had left records to the hi-fi magazines and record guides, a testament to what Nicholas Cook dubs "Plato's curse," or the priority accorded to written and printed sources within musical academia.[24] As Robert Philip recalls, this held true even when the subject of inquiry was performance itself: "Researchers were confident in recreating the unknowable sound of an early performance of *Messiah*, and yet dismissed the actual sounds of Elgar's own performances [of his works] as irrelevant (if they even knew of their existence)."[25] Recordings lacked credibility in the classroom, too: "The increasing importance of record collections was seen as a thoroughly undesirable consequence of incoming students' increasing inability to read scores," writes Daniel Leech-Wilkinson of England in the 1980s.[26] The landscape shifted with Philip's pioneering *Early Recordings and Musical Style*, whose publication in 1992 inaugurated a library of recording analysis that has grown steadily ever since.[27] Not every contributor challenges *Werktreue* explicitly, but the field's signature approach, comparing different recordings of the same music, does so by default. To cite a much-studied case: either a lot of distinguished pianists are mistaken, or there is more than one right way to play a Chopin mazurka.[28]

Opera scholars have come to similar conclusions. A transformation of opera staging in the 1970s and 1980s, especially by directors from spoken theater and film, and the concurrent rise of opera video inspired a surge of publications both on *Regieoper* ("director's opera") and on the mediating effects of film and television. Writers still worried about fidelity, not only of stagings to works, but now also of screen to stage. Yet here a new challenge emerged, for in their vividness and permanence, videos seemed like primary rather than secondary texts. Remembering their impact in the classroom, Marcia Citron writes: "Opera videos soon became much more than representatives of the work. Their interpretations took on a life of their own and generated thoughtful questions. What happened to the opera when it was adapted for the screen, and were these legitimate or successful changes?"[29] Together with several contemporary studies, Citron's *Opera on Screen* of 2000 shifts attention from composers and librettists to stage and video directors, to the conventions of theater and film, and to the social issues raised by many *Regieoper* productions, notably surrounding class and gender.[30] The interpreters become the interpreted, and the original sources serve as a background rather than an ideal.

In the 1990s and 2000s, the study of opera productions got another boost from a growing interest in performance, writ large, across many fields of the arts, humanities, and social sciences. Two books first published in 1999 have been particularly significant: Philip Auslander's *Liveness: Performance in a Mediatized Culture* and Hans-Thies Lehmann's *Postdramatisches Theater* (translated in 2006 as *Postdramatic Theatre*). Auslander argued that the categories of "live" and "mediated" had become so thoroughly intertwined in modern stage performance that the distinction between them was blurring, a claim with ramifications for both staged and recorded opera. Subsequent writing on opera video has treated "fidelity to the stage" as a chimera, indeed a distraction from the ways in which electronic media shape our perceptions even of live performance.[31] For its part, the "postdramatic" of Lehmann's title connotes a post-1960s, director-led turn from text toward performance, from telling stories that take place in other times and places—as in the Aristotelian concept of drama—toward creating experiences in the here and now. His focus is experimental spoken theater, but the protagonists include many of the same figures who populate *Regieoper* as well as revisionist productions of classic plays. Hence both David Levin, writing on opera, and William B. Worthen, the Shakespeare scholar quoted above, invoke Lehmann in their interpretations of radical restagings.[32] Regarding production "as a complex and autonomous (rather than merely derivative or decorative) expressive form," Levin explores its ability to do the work of his own paradoxical title, *Unsettling Opera*.[33]

The field of opera studies has dethroned the "work" in other ways as well. Research on adaptations signals a new respect for the numerous artists, beyond the original creators, who contribute to the life of an opera, whether it is the translators and choreographers who dreamed up the 1834 *Don Giovanni* with its fantastic ballets, or the prima donnas who intervened in many nineteenth- and early twentieth-century operatic texts.[34] In a different vein, Roger Parker revisits some celebrated instances in which composers themselves changed their minds, often to the dismay of critics and scholars. Among his examples is a pair of replacement arias written by Mozart for *The Marriage of Figaro*, vehicles for a new singer that have provoked much the same disdain as the Vienna additions to *Don Giovanni*. Parker disagrees, arguing that the arias have their own dramatic and musical logic, creating a new work rather than despoiling the original.[35] Carolyn Abbate, finally, stands *Treue* on its head in her aptly titled *In Search of Opera*, another catalyst for studies of performance since its publication in 2001. If works exist at all, in her view, they arise from events rather than texts: performances are not "ephemeral reflections of musical works," but their very source.[36] She evokes metaphors emphasizing immateriality, such as "souvenir": "Not a postcard or a piece of porcelain, nor a program book with color pictures and a cast list: [works] are not tangible, and you cannot put them in a drawer. Yet perhaps musical works, imagined objects, are mementos nonetheless, a reminder (though, unlike porcelain, not evident to the senses) that can stand for an embodiment, and what was once experienced in present time."[37] We carry operas in our heads, not our hands, like memories of favorite persons or places. Their shapes depend on things we have seen and heard—"the drama in its *doing*," to quote Worthen again—as well as the process of selecting and forgetting that goes into the formation of all memories.[38] Operas are doubly unstable, subject to variation not only in performance but in our own minds.

Clouds

In a later essay decrying musical scholarship's emphasis on text, Abbate targets recordings as part of the problem. They are just more tangible objects, ephemera made concrete, "tactile monuments in music's necropolis."[39] And indeed, in many ways opera recordings preserve even more fixed and invariant "works" than scores and libretti do. Unlike live performances, they endure as documents as well as memories, and by filling in so much of the detail that written sources lack—staging, tempo, sound color—they appear to leave less to the imagination. They also have their own authors, usually conductors or stage directors, whose names may appear in the same size font as the composer's (or bigger), and whose styles can be shown to develop much as musical styles do, at least as narrated in traditional histories of composition. Like Abbate's pieces of porcelain, finally, recordings can be put in drawers, or more typically on shelves—or at least, they could until streaming came along. As chapter 5 shows, the industry long tried to make opera recordings seem like books, especially boxed, multidisc sets of complete works. Tangible, covered with text and images, they are objects of culture to be catalogued and preserved.

Taken off the shelf and played, however, recordings are really no more static than books or scores. They activate the imagination without circumscribing its limits as much as it might appear. For all the information they contain, they leave out more: in audio recordings, the entire realm of the visual goes missing, and in video, we see only those visual dimensions that the cameras capture. Equally important, recordings are not experienced in a vacuum. Even those listeners or viewers who encounter only one *Don Giovanni* will take away different impressions depending on what they know of the subject matter, or of Mozart, or of opera generally. Play them a second recording, and the fixity of the details will expose the instability of the whole. The more variables are pinned down, the more obvious it is that no two versions are the same. No less than live performances, recordings are the grist for making operatic souvenirs.

This holds true even for recordings considered exemplary—"works" in their own right. Captured at the Salzburg Festival of 1954, one of Wilhelm Furtwängler's *Don Giovanni* performances has been issued and reissued several times, becoming something of a standard of comparison for other versions. Put in context, however, its apparent stability wavers. The culminating scene for Don Giovanni and the statue is famously epic, a consequence of the grand sonorities of the Vienna Philharmonic, the booming basses of Cesare Siepi and Deszö Ernster, and the glacial tread of Furtwängler's tempos. In the same year, the conductor would write of *Don Giovanni:* "A tempo is not a question of taste, but the natural expression of a particular work. Furthermore, the tempo in many cases is ductile, elastic."[40] The second sentence testifies to Furtwängler's penchant for rubato, which led Taruskin to cast him as the antimodern foil to the rhythmically regular modernists. But the first sentence is even more striking given the conductor's history with the opera. In a pair of less celebrated Salzburg recordings, from 1950 and 1953, he takes the same passage at a considerably faster tempo, about 76–77 beats per minute (bpm) to the quarter note in the opening Andante, as opposed to circa 60 bpm in 1954. Evidently the "natural expression of a particular work" can change.[41] ♫

All performers evolve, and even the modernist Stravinsky made three quite different recordings of *The Rite of Spring.*[42] But Furtwängler's conducting of the *Don Giovanni* overture suggests that something else is going on as well. The beginning of the overture prefigures the Andante of the statue scene, and in 1954 the two passages unfold at the same slow tempo. In 1950 and 1953, by contrast, where the statue scenes are faster, the overtures remain at slower speeds of approximately 55–60 bpm. Perhaps Furtwängler changed his mind about the relationship between the beginning of the opera and the end—or perhaps he just listened to his soloists. The Don Giovanni of 1950 was Tito Gobbi, whose voice was lighter than Siepi's and conception of the role more energetic. The statue of 1953, meanwhile, was Raffaele Ariè, a veteran of the festival at Aix-en-Provence, where one of Furtwängler's more modernist contemporaries, Hans Rosbaud, conducted Mozart's operas with comparative regularity and briskness. Formidable artists in their own right, the singers may have inspired the conductor to conceive the finale differently, such that the 1954 recording seems less like a definitive version than just another variation. More than that, it reminds us that no one creates opera alone, not even legendary conductors. Recordings provide ample evidence, partly by capturing the interactions between performers, partly by listing the dozens of contributors who are necessary to make opera happen. Some names may be in bigger type than others, but as they fill up the liner notes or the credits, they spread the creative responsibility from one to many.

Recordings mark the fluidity of works still further by putting everything in the present tense. Although each era has its distinctive styles and technologies, the whole point of sound and video recordings is to make live events available across time and space. Experienced en masse, they leave memories that are spatial as much as linear: watch or listen to enough recorded versions of *Don Giovanni*, even in chronological order, and what starts as a journey through the years can easily dissolve into a swarm of undated moments. I suspect this has long been the case; during the first half of the twentieth century, record enthusiasts listened to opera mainly in three- and four-minute snippets divorced from surrounding context, date of recording included. Digitization has magnified the effect, allowing entire archives to be navigated with ease, every *Don Giovanni* on record to be unmoored from histories and hierarchies. My concern—some might say obsession—with a single opera has undoubtedly heightened my sense of disorientation, and I am not suggesting that digitization leaves us in some kind of posthistorical fog. A few seconds of grinding shellac or glimpses of a bygone hairstyle will quickly dispel the mists. Nevertheless, anyone with a computer and an interest in performance knows the feeling. Thousands upon thousands of moments are out there, enduring, waiting to be experienced and reexperienced in any conceivable order.

When the *New Yorker* critic Alex Ross loaded the entirety of Mozart's music onto a portable digital audio player and listened to it over the course of three months in 2006 (speaking of obsession), he described what he heard as "the storm of style." Much of his account is chronological, going from the earliest works to the latest, but he also notes how the technology makes available "cross-sections of Mozart": "a dreamworld of Adagios; a neo-Baroque swirl of fantasias and fugues; a nonet of Quintets."[43] An archive of *Don Giovanni*s allows for comparable sampling, and were I to hazard my own metaphor, I would stick with Ross's but dial back the threat. The electronic *Don Giovanni* is like a cloud—a clichéd image currently, but no less apt for that. Musical or otherwise, masses of digital data are by nature amorphous, ever shifting, accessible from multiple angles. They do not set objects in front of us so much as envelop us in varied repetitions, so many that boundaries blur: where does one data set stop and the other begin? A digital archive gives us not just "*Don Giovanni* through the ages," but also "*Don Giovanni*, all the ages, all the time." The history of interpretation, the very subject that recording makes accessible, is forever collapsing into an eternal now.

Myths

All the more reason, then, to balance chronological narratives of *Don Giovanni* performance with more thematic inquiries. I have grouped the following chapters into three parts, each one treating a common recording format: excerpt audio recordings, complete audio recordings, and videos. To the extent that they dominate successive eras—respectively, the first and second halves of the twentieth century and the first two decades of the twenty-first—moving from one format to the next tells a story of long-range evolution. At the same time, the breadth of each era leaves ample room for exploring the variety within it and its connections to the others, the resonances that cross time and space. To this end, while each part begins with a chapter on the format itself, its conventions and their consequences for how *Don Giovanni* looks and sounds in a given period, the remaining chapters explore both linear and nonlinear relationships between select examples. Recordings separated by decades may have as much in common as those made in the same year.

My choice of examples is informed by two substantial bodies of literature. The first concerns recording, which has inspired torrents of commentary ranging from reviews of individual releases to reflections on how discs or tapes or digital files revolutionize operatic experience. In addition to a wealth of critical insight, these texts convey vital information about performance traditions and technological innovations dating back to the nineteenth century. The second body of literature, concerning *Don Giovanni*, offers a kind of counterpoint, a trove of interpretation often inspired by performance but intended to illuminate the "work." It includes some of the best-known writings on any opera, notably a rhapsodic short story from 1813 by E. T. A. Hoffmann, "Don Juan: A Fabulous Event That Happened to a Traveling Enthusiast," and a set of philosophical speculations by Søren Kierkegaard incorporated into his *Either/Or* of 1843. Together they helped originate several resilient themes, including the claim mentioned above that libertine conquest is an expression of forces transcending ordinary human experience or morality. Yet theirs are but two voices in a conversation that also includes Victorians dubious of the opera's moral framework, Freudians preoccupied with its oedipal implications, modernists interested in its "purely musical" structure, and feminists critical of its misogynist legacy. Whether reflective of performance or reflected by it, their commentaries highlight the social and musical questions that make the opera such a problematic centerpiece of the repertoire.

In part I, concerning excerpt recordings, I begin from the vantage point of character. Wax cylinders and 78 rpm discs mostly preserve arias and small ensembles, and though the industry made some effort to acquaint buyers with the parent operas, marketers and critics focused on star singers and their ability to turn three- to four-and-a-half-minute records into miniature theaters of feeling. As chapter 1 illustrates, in this context *Don Giovanni* was especially valued for tuneful numbers that suited the format and the thirst for romantic outpouring. Chapters 2–4 look at several of the most popular examples, recordings of which reveal both a broad revolution in style and a more individual, sometimes fractious contest over the nature of the opera's characters. As Nicholas Cook and others demonstrate, the earliest generations of singers caught on record favored flexible speeds and lots of ornamentation, shaping each phrase with what Cook calls a "rhetorical" attention to detail.[44] By contrast, singers who came of age after 1900 tended toward steadier tempos and cleaner lines, suggesting the modernist regularity advocated by Stravinsky and others. *Don Giovanni* performance changes in concert with the generations, becoming generally less heated and more suave, less "in-the-moment" and more reflective. At the same time, singers find different meanings in the same musical numbers, not only across generations but within them. Recordings of Don Giovanni's "Deh vieni alla finestra," often referred to as the Serenade, reference a litany of strategies favored by historical libertines, while those of "Fin ch'han dal vino," the so-called Champagne Aria, and of the solo turns by Don Ottavio and Zerlina project competing visions of masculinity and femininity. A favorite excerpt may fall anywhere on a continuum from flattery to force, virility to sentimentality, naïveté to transcendence.

Part II examines complete audio recordings from a second vantage point: time. Allowing listeners to hear multiple numbers in sequence, LPs and CDs emphasize unfolding events and interactions as much as they do expressive moments. As chapter 5 shows, they also document the later twentieth-century reckoning with *Don Giovanni*'s subject matter. Even as complete recordings strengthened the opera's status as a cultural monument, they brought new attention to its most problematic passages, in part by making them available for repeated listening, in part by including translations, images, commentaries, and other explanatory material. Chapters 6 and 7 consider two particularly troubling scenes: the opening of act 2, in which the protagonist "reseduces" his abandoned spouse Donna Elvira, and the act 1 finale, where he tries and fails to rape Zerlina. Complete recordings capture a second revolution in performance style, in which the steady rhythms, expansive sonorities, and long lines of modernism give way to the even steadier rhythms, downsized forces, and articulated phrases of historically informed performance. In addition, Mozart's secco recitatives (those accompanied by keyboard or continuo ensemble rather than orchestra) go from being terse bridges between concerted numbers to eventful dialogues in their own right. At the beginning of act 2, all of this tempers the passage's claims to comedy: as the concerted numbers become quicker and more down-to-earth, and the recitatives slower and less slapstick, the cruelty of Donna Elvira's treatment becomes all the more apparent. In the act 1 finale, a sequence of concerted sections unbroken by recitative, the changes rebalance the relative agency of the characters. With the slower sections growing swifter over time and the trajectory of tempos less extreme, the focus shifts from the protagonist's fabled energies to the others' efforts at resistance.

Part III revisits questions about time and character within the physical and mental spaces opened up by video. Chapter 8 explores how the filmmaker's vocabulary of cuts, close-ups, and zooms shapes viewers' perceptions, especially of psychological interiority. Chapters 9 and 10 then turn to the most frequently analyzed psyches in the opera, those of Don Giovanni and his chief antagonist, Donna Anna. Video recordings preserve still more wide-ranging developments in performance, notably the post-1970s rise of updated settings and costumes and the concurrent shift from a comparatively static and decorous manner of acting to a more dynamic and naturalistic one. In Donna Anna's case, the transformation revives a question posed by nineteenth-century critics: is her character driven by hatred of Don Giovanni, or by guilt and hidden desire? Productions issued on video have increasingly favored the latter, putting new emphasis on Donna Anna's inner life while also suggesting, intentionally or not, that split loyalties are more psychologically compelling in a woman than trauma. As for Don Giovanni, over time he has looked more and more traumatized himself, or at least buffeted by the social ills of authoritarianism, exploitative capitalism, or nuclear-family repression, depending on the production. The opera is no longer about him so much as the inequities and neuroses that produce libertines, for which everyone is to blame. Rake, predator, or neurotic, the protagonist operates in a world in which moral compromise and ambivalence are universal.

So serious has *Don Giovanni* become that I sometimes wonder if the myth has run its course. Not every recent production critiques society writ large, but in today's theaters, even tights-and-swords versions may reject old-school hijinks in favor of the "musical realization of a monstrous world and its deliverance," as Richard Brody described a 2015 performance at the Metropolitan Opera in New York.[45] The cast comes together, especially the women, to resist a powerful man's aggression and eventually celebrate his downfall. Finishing this book in the era of #MeToo and the exposure of so many real-life Don Juans in opera and elsewhere, I find it difficult to imagine the forceful Eros of yesteryear returning unchallenged to the stage, or even the more genial "endearing swashbuckler-sensualist" whom David Littlejohn saw in the mid-twentieth-century *Don Giovanni* of Ezio Pinza.[46]

On the other hand, quite apart from the fact that recordings keep older visions of the opera alive, latter-day productions are spawning myths of their own, substituting psychology for morality and social circumstance for personal guilt. The libertine can still emerge a hero, and he will no doubt continue to do so until gender roles change more fundamentally than they have to date, not least within opera performance and criticism. The statistics are sobering. Of the nearly sixty *Don Giovanni*s I have collected on video, only three have women stage directors,[47] and to my knowledge only a single woman conductor appears in any commercially released recording, audio or video.[48] Women figure more prominently as commentators, especially in recent years, and of course numerous women artists other than directors and conductors, singers above all, have contributed substantially to the opera's twentieth- and twenty-first-century remaking—as have plenty of men. Yet it takes nothing away from their efforts to suggest that a greater presence of women in leadership roles, whether as creators or critics, will bring about a further reckoning with the drama and its legacy. My account aims to provide perspective and, perhaps, inspiration. *Don Giovanni* is no longer the opera it was in 1787, or even 1987, and its journey thus far merits the attention of performers and critics, teachers and students, audience members and record collectors alike. What happens next is up to us.

PART I
Clouds of Feeling

Excerpt Audio Recordings

CHAPTER ONE

Imagining Excerpts

The early recording industry loved the opera excerpt. Companies relied on opera to legitimate their enterprise, pitching records and playback equipment as a means of accessing high culture. Advertisements showed living rooms filled with star soloists, and to judge from the records, their "performances" were mostly of songs.[1] The dominant formats of the era, the "short-playing" wax cylinder and the shellac 78 rpm disc, delivered individual numbers more elegantly than whole dramas, and the brief, emotionally concentrated vignette proved enduringly popular. From the late nineteenth century, when recording emerged, through to the 1940s, longer-playing formats fell flat and complete opera recordings remained a rarity. Among them was a single complete *Don Giovanni*, issued on twenty-three double-sided discs in 1936.[2] Though an invaluable document, it barely registers against the several hundred excerpt recordings that were released for this opera alone.

Those recordings can tell us a lot about the performance of the parent opera. Mainly arias and small ensembles, they offer a kind of aural dramatis personae, a roster of characterizations undoubtedly reflective of stage practice, at least to an extent. The industry and the new field of record journalism promoted the connection, offering translations, synopses, and photographs to illuminate the original context of each selection. At the same time, excerpts have always done more than evoke complete operas, whether the printed numbers that have circulated since the earliest days of the genre or the recorded "highlights" that appear to this day. They bring to mind performers, occasions, places, moods, musical styles—a whole world of associations that may or may not have anything to do with a given opera, or indeed any opera. Where excerpts on record are concerned, consumers who were not operagoers (and many who were, for that matter) may well have heard *Don Giovanni* discs in terms of other kinds of music, or experiences surrounding the buying and playing of records themselves. As much as they preserve operatic performance practice, excerpt recordings also bear witness to a broader culture of lyrical expression and exchange, most of it vocal, and all of it mediated by evolving technologies and markets.

How to Read a Record

My own introduction to short-playing opera dates to the same period in the early 1990s when I was first trying to teach *Don Giovanni* (see the introduction). In the midst of debating which version to play for my students, I discovered the Offenbacher Mozart Collection at the University of Washington, an archive of over fifteen hundred cylinders and 78 rpm discs of the composer's vocal music.[3] Encompassing well-known recordings that have appeared in LP and CD transfers, along with many others that have not, it opened my ears to a diversity of interpretation no less striking (or daunting) than what I encountered in the full-length audio and video releases of the Mozart bicentennial years. This invaluable resource forms the basis for part I.

The contents of the collection's 370 *Don Giovanni* recordings reveal something of the industry's priorities. They lean heavily toward certain numbers, notably the arias of Don Giovanni, Don Ottavio, Leporello, and Zerlina, as well as the duettino for Don Giovanni and Zerlina, "Là ci darem la mano" (fig. 1.1). Donna Anna's and Donna Elvira's solos appear much less frequently, and other ensembles even less so. The proportions may differ somewhat from those found across all *Don Giovanni* releases of the era,[4] but the chosen numbers suit both the general practices of the format and some specific preferences regarding this opera. They have much in common with the ballads and popular songs that dominate early recordings in general; in addition to being brief, they favor tuneful melodies, clear melody-and-accompaniment textures, and (mostly) straightforward romantic texts. They also support the industry's interest in emphasizing love and comedy over violence and supernaturalism. As we shall see, the marketing vision for *Don Giovanni* could better accommodate the suave romancing of the title character's Serenade, "Deh vieni alla finestra," or the antic boasting of Leporello's Catalogue Aria, "Madamina, il catalogo è questo," than it

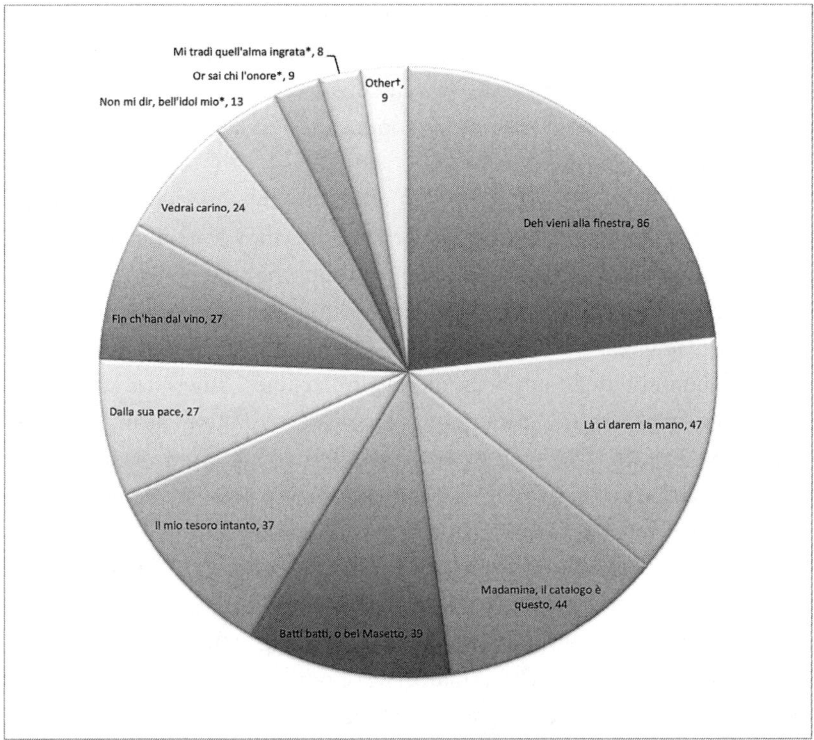

FIGURE 1.1 *Don Giovanni* in the Offenbacher Mozart Collection. Totals include multiple recordings by the same singer(s), but not multiple releases of the same recording.
* Includes versions with and without the foregoing recitatives.
† Includes the ensembles "Ah taci ingiusto core" (two recordings), "O statua gentilissima" (2), and "Eh via buffone" (1); Leporello's solo, "Notte e giorno faticar" (2); an excerpt of the act 1 finale, the menuetto (1); and an excerpt of the act 2 finale, "Già la mensa è preparata" (1).

could the rage and sorrow of the noblewomen's arias or the complexities of Mozart's ensembles.

The physical discs and cylinders suggest additional priorities, not all of them related to the parent opera. They came with comparatively little information, mere labels as opposed to the illustrated boxes and explanatory booklets that would accompany operas on LP and CD. Yet still they sought to direct listeners, especially discs, which proved the more enduring medium. Disc labels for opera, splashes of color against the solid background of the shellac, often signal the high-art pretensions of the contents (fig. 1.2). Several firms mixed modernism with neoclassicism, associating records with columned temples, lyres, and figures of myth (an angelic muse on Fonotipia's labels, Romulus and Remus on Cetra's). The images evoke cultivation and privilege, as does the frequent use of gold or silver lettering on a deep-colored background. The latter practice lends a rarified air even to Victor's labels, with their homely drawing of Nipper the dog, and indeed Victor and its corporate descendants would use color to signal exclusivity well into the LP era, issuing opera and classical music on a special "Red Label."

The texts of the disc labels convey more tangible information and still another set of priorities. The largest print names the company, drawing attention to its role as cultural curator. Next in prominence is most often the opera, sometimes set in large type or all caps (figs. 1.2a, 1.2b, and 1.2d). Companies may have thought the work titles would be more familiar than those of the individual numbers; along similar lines, where an aria or ensemble title appears first, it sometimes bears an attribution, as in "Serenata from Don Giovanni." To this extent, the labels do highlight the parent opera; to purchase an opera record was to buy into a story or a stage event.[5] On the other hand, it was also to invest in a singer, whose name and credentials generally overshadow those of accompanists or even composers. Many labels indicate voice types, and some identify cities or opera companies with which the soloists are associated (e.g., fig. 1.2a). By contrast, the composer's name may be set in small type or parentheses, and the accompaniment reduced to a generic "piano" or "orchestra" (figs. 1.2b and 1.2c)—if it is mentioned at all (fig. 1.2a). Just as the content of early recordings revolved around song, so performance did around singers, and everyone else remains subsidiary through much of the short-playing era. Record company, selection, and soloist were the items the industry considered essential, the initial stimulus to the listener's imagination.

Musical Companions

Programming combinations were a different kind of prompt. As short as they were, wax cylinders and the original single-sided discs sometimes included more than one selection: a 1909 Victor disc, featuring the baritone Antonio Scotti, follows Don Giovanni's "Deh vieni alla finestra" with the equally brief "Quand'ero paggio" from Verdi's *Falstaff.* Once the two-sided disc prevailed in the 1920s, nearly every selection came with a partner. Of course, the phonograph operator had the final say over what was heard and in what order, but two-sided discs invited consumers to think of multiple selections in relation to one another, selections that were usually performed by the same soloist.

Across the Offenbacher collection, *Don Giovanni* excerpts are most

FIGURE 1.2 78 rpm disc labels, 1907–55

a. Giuseppe Anselmi, "Il mio tesoro intanto." Fonotipia 62167, 1907. Courtesy Music Library Special Collections, University of Washington Libraries. Photograph by the author.

b. Elisabeth Schumann, "Schmäle, schmäle, lieber Junge" ("Batti batti, o bel Masetto"). Odeon 76726, 1917. Courtesy Music Library Special Collections, University of Washington Libraries. Photograph by the author.

c. Ezio Pinza, "Fin ch'han dal vino." Victor 1467, 1930. Courtesy Music Library Special Collections, University of Washington Libraries. Photograph by the author.

d. Giuseppe Taddei, "Deh vieni alla finestra." Cetra AT 0403, 1955. Courtesy Music Library Special Collections, University of Washington Libraries. Photograph by the author.

commonly paired with other music from *Don Giovanni*, especially other numbers for the same characters. About 20 percent of the recordings of "Deh vieni alla finestra" appear together with renditions of "Là ci darem la mano" or of Don Giovanni's Champagne Aria, "Fin ch'han dal vino," and roughly a third of recordings of Zerlina's "Batti batti, o bel Masetto" and Don Ottavio's "Il mio tesoro intanto" flip over to play the same characters' "Vedrai carino" and "Dalla sua pace" (the other popular number, Leporello's "Madamina, il catalogo è questo," generally takes up both sides of a two-sided disc and thus appears by itself). Whether or not they were heard in sequence, two excerpts from the same opera point again toward the parent work, and two arias by the same character suggest a kind of profile, sometimes reinforced by attributions. When Maria Cebotari recorded "Batti batti, o bel Masetto" and "Vedrai carino" in 1941, Deutsche Grammophon subtitled each one "Arie der Zerlina," as if they were elements in a characterization.

When *Don Giovanni* is matched with other sources, the choices differ by character. The arias of Zerlina and Don Ottavio tend to appear with like examples. The most common partner for "Batti batti, o bel Masetto" is Susanna's "Deh vieni non tardar" from Mozart's *The Marriage of Figaro*, another pastoral love song for a buffa heroine; and for "Il mio tesoro intanto," it is Tamino's "Dies Bildnis ist bezaubernd schön" from *The Magic Flute*, another romantic effusion for an idealistic aristocrat (tables 1.1 and 1.2). Both choices point up affinities between characters in different operas, as do, for the most part, discs featuring music by other composers. The non-Mozart selections paired with Zerlina's and Don Ottavio's arias all have romantic overtones of some kind, and in the broadest terms they resemble Mozart musically; apart from Berlioz, Cilea, and Puccini, the composers in

TABLE 1.1. Selections paired on disc with "Batti batti, o bel Masetto"

COMPOSER	OPERA	SELECTION
Cilea	*Adriana Lecouvreur*	"Io son l'umile ancella"
Donizetti	*Don Pasquale*	"La morale in tutto questo"
Gounod	*Faust*	"Faites-lui mes aveux"
Martini	n/a	"Plaisir d'amour"
Mozart	*Don Giovanni*	"Vedrai carino" (13)
		"Mi tradì quell'alma ingrata"
	The Marriage of Figaro	"Deh vieni, non tardar" (5)
		"Voi che sapete" (2)
	Il re pastore	"L'amerò sarò costante"
	The Magic Flute	"Bei Männern, welche Liebe fühlen"
		"Ach ich fühl's"
Puccini	*Tosca*	"Vissi d'arte"
Rossini	*Semiramide*	Bel raggio lusinghier"

TABLE 1.2. Selections paired on disc with "Il mio tesoro intanto"

COMPOSER	OPERA	SELECTION
Berlioz	*La damnation de Faust*	"Nature immense"
Donizetti	*L'elisir d'amore*	"Una furtiva lagrima" (2)
		"Quanto è bella!"
Flotow	*Martha*	"Ach, so fromm"
Gounod	*Faust*	"Salut, demeure chaste et pure" (2)
Handel	*Semele*	"Where'er you walk"
Mozart	*Così fan tutte*	"Un'aura amorosa"
	Don Giovanni	"Dalla sua pace" (14)
	The Abduction from the Seraglio	"Wenn der Freude Tränen fließen"
	The Magic Flute	"Dies Bildnis ist bezaubernd schön" (4)
Puccini	*La bohème*	"Che gelida manina"
Thomas	*Mignon*	"Elle ne croyait pas"

tables 1.1 and 1.2 all preserve late eighteenth-century conventions of periodic phrasing, formal repetitions, and lucid melody-and-accompaniment textures. The commonalities highlight stock character types spanning genres and eras, the soubrette or the sentimental tenor.

Discs of Don Giovanni's numbers have more variety, especially discs featuring "Deh vieni alla finestra." Its sheer popularity led to its being paired with dozens of sources, from *The Marriage of Figaro* and *The Magic Flute* all the way to Erich Wolfgang Korngold's *Die tote Stadt* of 1920. Many of the selections result in more contrast than continuity, recalling the original inasmuch as Don Giovanni's other solo in the opera itself, and the most frequent companion of "Deh vieni alla finestra" on disc, is the frenetic "Fin ch'han dal vino" (fifteen discs). Several other choices engineer a comparable shift in mood, among them Leporello's "Madamina, il catalogo è questo" (one disc), the Count's "Vedrò mentre io sospiro" from *Figaro* (one disc), and the "Chanson du toréador" from Bizet's *Carmen* (three discs). All three set the elegant persuasions of the Serenade against more ominous impulses, whether sexual boasting, jealous bombast, or masculinist grandstanding. By virtue of association, the aria may sound a little less appealing, more like the dangerous deception that it actually is in the story. More comedic choices, such as Papageno's "Der Vogelfänger bin ich ja" (two discs) or Falstaff's "Quand'ero paggio" (seven discs) have a different effect: though also full of sexual overtones, their characters' boasting is more pathetic than threatening, and Papageno's number has a folksiness

TABLE 1.3. Nonoperatic selections paired on disc with "Deh vieni alla finestra"

COMPOSER	SELECTION
Anonymous	"Margoton va t'à l'eau"
Jean-Paul-Égide Martini	"Plaisir d'amour"
Augusto Rotoli	"Mia sposa sarà la mia bandiera"
Johann Strauss II	"Frühlingsstimmen"
Francesco Paolo Tosti	"Ninon"

that draws out the humbler associations of the Serenade's tuneful melody and gentle 6/8 rhythm. Such discs seem comparatively lighthearted, free of brooding violence or upper-class pretension.

Yet another standard prevails on a handful of discs matching "Deh vieni alla finestra" with nonoperatic selections (table 1.3). Here the choices are all ballads or love songs, ranging from an arrangement of a French folk song, "Margoton va à l'eau," to an eighteenth-century apostrophe to love, Jean-Paul-Égide Martini's "Plaisir d'amour," to a nineteenth-century sentimental farewell, Augusto Rotoli's "Mia sposa sarà la mia bandiera." Beyond subject matter and lyricism, all share the clear textures and strophic song form of "Deh vieni alla finestra"—and, in the case of "Plaisir d'amour," its swaying 6/8 rhythm. They draw Mozart's number into an orbit of romantic crooning typified by Martini's lament, which was ubiquitous on both opera and popular music records and would become, in adapted form, a famously sentimental hit for Elvis Presley, "Can't Help Falling in Love" (1961). Heard in its company, the Serenade makes its own claim to be a stand-alone, timeless love song.

Educating Listeners

Whatever appeared on the discs, additional resources stood ready to guide listeners' interpretations. Record advertising and media coverage were organized around singers as much as composers or works, and a soloist's public persona could provide suggestive clues. An outstanding example is the American soprano Geraldine Farrar, who appeared in films as well as onstage and was a regular subject of profiles, interviews, and gossip. She modeled a strong and quintessentially modern femininity, one sign of which was her reimagining of familiar operatic heroines. In 1911 New York's *Theatre Magazine* published an interview and photo essay exploring her new costume designs for half a dozen roles, Zerlina among them. Though she details historical and dramatic reasons for each choice, the common denominator is unmistakable: in side-by-side images, every revision looks less confining, more flowing, and in Zerlina's case more exotic than the original (fig. 1.3).[6] The designs are some distance from the flapper look that Farrar would champion in the 1920s, but all relax the prim starchiness of Victorian fashion. Listeners could easily have brought the spirit of such images—lively, assertive, up-to-date—to their experience of her numerous recordings, and U.S. consumers may even have felt patriotic stirrings. "Miss Farrar's story is one that the ranchman or miner can understand," wrote Willa Cather in 1913. "It gratifies his national pride, meets his sense of the picturesque. When he puts a 'Farrar record' into his phonograph, he has something of the feeling of part-ownership."[7] Along with everything else, her voice could be a sign of American high-cultural bootstrapping.

Record companies also undertook to educate consumers more directly, using various means to circulate the kind of information that would later appear in LP and CD booklets. The *Edison Phonograph Monthly*, a magazine targeted at retailers that was published from 1903 to 1916, offered talking points for each of the firm's new recordings. For its "Grand Opera" series of 1910–11, these included synopses of the excerpts and, in some cases, English translations, all to be shared with prospective customers.[8] Edison also experimented with recorded program notes. On the reverse side of a "Batti batti, o bel Masetto" disc from 1917, an "Explanatory Talk" offers a history of the opera, a précis of the dramatic situation, a translation of the aria, and a fawning introduction of the soloist, Marie Sundelius (fig. 1.4). Rolling his r's assiduously, the lecturer, Harry E. Humphrey, assures the listener of the material's European high-art origins, and the lecture itself introduces some key assumptions of *Don Giovanni* commentary of the time. Humphrey avoids anything that might offend contemporary standards of respectability, limiting his synopsis to only two sentences: "Zerlina is betrothed to Masetto, a peasant. Masetto is provoked at her, and in this aria she begs his forgiveness." Notwithstanding an earlier reference to "the licentious young nobleman," one would never know that Don Giovanni was the cause of Masetto's provocation, or Zerlina's distress. Nor does the translation read by Humphrey let on that the original Italian, duly sung by Sundelius, is laden with titillating imagery. While translations are never wholly neutral—remember the video subtitles quoted in the introduction—degrees of poetic license differ widely. Comparatively literal renderings of "Batti batti, o bel Masetto," such as that found in the 1990 libretto by David Stivender and Susan Webb, preserve the violence of opening line and the

MISS FARRAR IN THE ZERLINA COSTUME OF TRADITION

MISS FARRAR IN HER NEW ZERLINA COSTUME

all are composites, authentic in detail, of that same Italian renaissance in which period Juliet lived.

"In the ballroom scene I discarded the conventional white, in favor of a silver embroidered, delicate pink, almost colorless, because it floated better, and appeared more aerial, being less affected in the folds by the stage lights, which would have given to pure white a greater heaviness.

"To appear on the balcony in ball costume is for that scene traditional. Instead, I wear a little slip and veil. Romeo, himself, in the preceding scene below her windows, gives the impression that Juliet is disrobing. It is, too, a natural assumption that such would be done before she steps upon the terrace for a breath of air. One must be guided by the intelligence.

"Traditions are made to be broken—by the few. Memory is the enemy of the present. Because things have gone on in one groove, a stilted, inaccurate, maybe farcical one for years, is no reason that they should be continued so and not altered now. There are those who lie awake at night to conjure up something unusual and eccentric. But that is not making new traditions, which mean instead a supplying of something more correctly authentic than the ones existing. A grafter in music, a grafter in anything, is despicable. And that is exactly what blind, unquestioning acceptance of operatic tradition in dress means.

"Take Manon's costume in the last act of Massenet's 'Manon,' for twenty years it has consisted of a uniform and with cut hair, both to denote the convict condemned to a penal settlement. There has been no authority for prison garb. People of all conditions, fine ladies, gallants, peasants, were taken as they were. Knowing that the real Manon would have worn the dress common to her day, I have cut away from tradition, which had no foundation, and costumed the part according to truth.

"The selection of stuffs is, perhaps, the most difficult branch of costuming to decide. Take those used for the Goose Girl in 'Königskinder,' as example. For weeks I experimented with cashmere, woolens, denham, a mass of fabrics, to secure the right effect, to make the character individually correct, of just the right appeal to the eye as a true picture. She would have to appear fourteen or fifteen, free, slender, supple, not held in by stays. To find materials that will keep in shape, yet fall into proper lines in these conditions, is more difficult than in the case of more elaborate costumes.

"Other Goose Girls will rise up, not fifty of them, but a favored few, who will break the

MISS FARRAR WEARING THE MANON COSTUME OF TRADITION

MISS FARRAR AS SHE THINKS THE PART SHOULD BE DRESSED. NEW COSTUME OF MANON

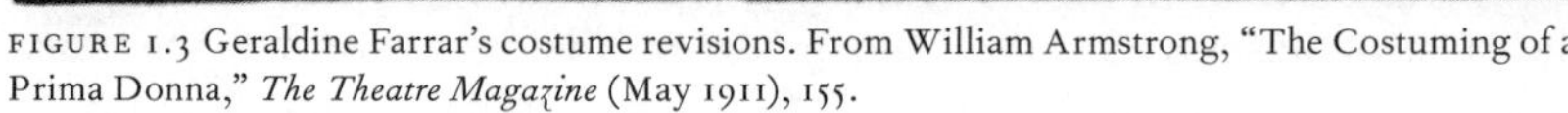

FIGURE 1.3 Geraldine Farrar's costume revisions. From William Armstrong, "The Costuming of a Prima Donna," *The Theatre Magazine* (May 1911), 155.

FIGURE 1.4 Harry E. Humphrey, "Explanatory Talk" accompanying Marie Sundelius, "Batti batti, o bel Masetto." Edison 82120, 1917. Courtesy Music Library Special Collections, University of Washington Libraries. Photograph by the author.

suggestiveness of the conclusion, which refers to "night and day" pleasures. The version used by Humphrey does not; written by Isaac Pocock for an 1817 adaptation of the opera by Henry Bishop titled *The Libertine*, it substitutes scolding for beating and devotion for delight:

1787:	Batti batti, o bel Masetto, la tua povera Zerlina:
1990:	Beat, beat your poor Zerlina, oh handsome Masetto
1817:	Chide me, chide me, dear Masetto, chide Zerlina at your will
[. . .]	
1787:	in contenti ed allegria notte e dì vogliam passar
1990:	in delights and gladness let us pass the night and day
1817:	Night and day to be devoted here I vow to err no more[9]

Pocock's words help Humphrey reduce a tangle of sexual tensions to a simple reconciliation, something he can call "a beautiful little aria with all the daintiness and graceful charm that only Mozart could produce."

Victor's Red Label series gave rise to a similar but more enduring resource than Edison's efforts: *The Victor Book of the Opera*, first published in 1912 and updated regularly thereafter. It claimed to be much more than a record catalogue. "Even though the scenery and costumes may be lacking," it says in the foreword, "the absence of these accessories will now be atoned for in some measure by the graphic descriptions and numerous illustrations in this book."[10] On the opening pages, photographs of famous opera houses (first exterior, then interior) usher consumers "inside," much like the scene-setting shots of later opera videos, and there follows a series of chapters on individual operas, with further photographs depicting the recording artists in character. The accompanying text includes synopses, dramatis personae, histories of composition, and translations of the numbers available for purchase.

The pages on *Don Giovanni* evince the same cheeriness as Humphrey's explanatory lecture, never acknowledging an underside to the protagonist's "gay conquests," as they are called here. Both "Batti batti, o bel Masetto" and the episode that motivates it, Don Giovanni's seduction of Zerlina in the duettino, are presented as lighthearted occasions for pretty music: "The *Don* then proceeds to flatter the young girl and tells her she is too beautiful for such a clown as *Masetto*. She is impressed and coquettes with him in the melodious duet, *La ci darem* [*sic*], the witty phrases and delicate harmonies of which make it one of the gems of Mozart's opera . . . *Zerlina* is endeavoring to make her peace with *Masetto*, but he is sulky. She then sings her lovely *Batti, batti*."[11] Pocock's "Chide me, chide me" translation follows, and the surrounding photographs picture happy Zerlinas and dashing Dons. The advertised selections convey a similar message, comprising four recordings each of the duettino and of "Deh vieni alla finestra," two of "Madamina, il catalogo è questo," one of "Batti batti, o bel Masetto," and one of the menuetto from the act 1 finale (for orchestra only). Nothing from the noblewomen, nor from the more licentious Don Giovanni of "Fin ch'han dal vino." In *The Victor Book*, the opera is all fun and romance, entirely suitable for recreation in the home.

When *Gramophone* magazine began publication in 1923, it featured similar if less systematic aids for record fans, including translations of frequently recorded arias, synopses of the surrounding action, and occasional surveys of multiple recordings from the same opera.[12] *Gramophone* also documented another resource: British gramophone societies, groups of enthusiasts who gathered periodically to share records and show off playback equipment. Typically, one member presented a series of discs with commentary, nearly always a potpourri of arias, songs, or instrumental movements—a reminder of how frequently excerpts were heard and interpreted apart from their parent works. On the other hand, at least some "gramophone programs" appear to have offered a kind of oral equivalent to *The Victor Book*:

A large attendance in the Onward Hall, Manchester, on Monday, May 12, [1924,] greeted Miss G. Elsworth, of Lymm, the solitary lady member who ventures the ordeal of demonstration. . . . The subject, *Rigoletto*, was descriptively prepared and attractively delivered, fully revealing the dramatic subtleties of the *Jest* with its pathetic finale, and prefaced by the story of Verdi's life, proved singularly impressive. All the principal "Airs" were rendered by celebrities—Caruso, Battistini, Galli-Curci, Tetrazzini, Amato, Schipa, Homer, and De Luca—with that gripping and wonderful artistry which makes one ever grateful for the permanent opportunity of reviving those glorious vocal tones and interpretations through the gramophone, otherwise impossible to the financial means of the average Britisher, even were the artists available.[13]

If this sounds a little like a music-appreciation class, elsewhere the magazine does indeed proselytize for the educational potential of the technology. This was more formally realized in gramophone-illustrated lectures by music professionals and eventually by recorded anthologies of Western music history, most ambitiously the *History of Music in Sound*, issued by HMV near the end of the short-playing era to accompany *The New Oxford History of Music*.[14] Though not limited to opera, such scholarly enterprises offered additional guidance to listeners curious about the musical and historical background of what they heard on record.

Listening Companions

Listeners may have been further influenced by where and with whom they listened: obviously so, in lecture-presentations attended by opera and classical-music enthusiasts, but also in other situations, even alone at home. A phrase coined by Evan Eisenberg, "ceremonies of a solitary," captures a familiar mythology of listening to records, evoking the fan who consumes music in isolation, undisturbed by other audience members or distractions of any kind.[15] In Eisenberg's telling, the founder of *Gramophone*, Compton Mackenzie, took this experience as a kind of blueprint for life, seeking a permanent state of solitude by spending his time on privately owned islands. Where opera is concerned, solitary listening evokes an equally familiar image of vicarious participation,[16] the history of which dates at least as far back as Thomas Mann's novel *The Magic Mountain*, published in 1924. Near the end of the book, the sanatorium-bound hero, Hans Castorp, becomes enamored of a state-of-the-art player and contrives to spend many hours listening by himself, constructing elaborate fantasies.

Again it was French music [Bizet's *Carmen*]—an opera Hans Castorp knew well, having seen and heard it repeatedly. . . . The record took up the play at the second act, in the Spanish tavern, in crude Moorish architecture, a shawl-draped, roomy cellar like the floor of a barn. One heard Carmen's voice, a little brusque, yet warm, and very infectious in its folk-quality, saying she would dance before the sergeant; one heard the rattle of castanets. . . . Hans Castorp played this single record over and over, and listened with the deepest participation. As far as its contents went, it did not fetch the action much further; but its imploring emotion was moving in the highest degree.[17]

The passage (which includes an extensive synopsis as well) has much in common with the advertising blurbs quoted above. With *Carmen* as with *Don Giovanni*, records privilege feeling over action; the feelings remain respectable; and the imagined scenario involves a young woman performing for a man, much like Zerlina and with the same whiff of voyeurism. If men at the opera could take pleasure from watching women implore other characters, men at home could fancy that they were the ones being implored.

Records do not aid Castorp's cure. At a time when listeners were still getting used to hearing distant or deceased voices come out of a machine, gramophones could readily connote death,[18] and Mann pictures a morbid obsession intimating the hero's demise in World War I. He plays on other associations from the early years of recording as well, when the technology was understood both as an analogy for the human mind, particularly memory, and as a tool for accessing the unconscious.[19] Though his records preserve opera rather than his inner life, Castorp's "ceremonies" still seem like symptoms,[20] and for actual flesh-and-blood listeners, solitary record consumption may have had isolating if not pathological results. In the same year *The Magic Mountain* was published, a correspondent wrote to *Gramophone* that beating time to his favorite records was "infinitely more exhilarating than dancing."[21]

That said, not everyone listened alone, and not everyone who did fell into unhealthy introspection.[22] Recorded music was heard regularly in public or semipublic settings, such as the gramophone society meetings described above or "gramophone concerts" hosted by stores and civic institutions. Even in Castorp's sanatorium, the gramophone originally entertains groups of patients at once, with the hero serving as disc jockey. Records were played in restaurants, clubs, and brothels, not always to attentive listeners but with the express purpose of fostering sociability. If advertisements are any guide, families and friends listened together in the home, frequently enough that the domestic gramophone program became a subject of satire.[23] Even loners, furthermore, may have found themselves drawn into virtual communities, bound by what William Howland Kenney terms "circles of resonance"—"group sensibilities in which listeners shared, debated, analyzed, and fought, often passionately, over their personal patterns of empathy and appreciation for what they heard in the grooves of 78 rpm recordings."[24] As an example, he cites the collective nostalgia expressed by rural Americans who bought Edison's recordings of nineteenth-century parlor songs. Willa Cather imagined a similar sense of kinship among consumers of "Farrar records," and even the island-dwelling Compton Mackenzie fostered a community by providing a forum, through *Gramophone*, for the discussion of classical music and opera. His subscribers may not have been sitting together in an auditorium, but they shared an experience of music all the same.

Then as now, the community surrounding opera records had definite economic and social boundaries. Records were regularly promoted as making high culture more accessible, and when *The Victor Book* claimed that "for every person who can attend the opera there are a hundred who cannot," its editors may have been imagining Cather's "ranchman or miner" listening to Farrar, or the "average Britisher" experiencing *Rigoletto* at a gramophone society meeting.[25] Nonetheless, opera and classical music records cost more than their popular counterparts, and, judging again from advertisements, the target market either possessed or aspired to the accoutrements of middle- and upper-class life, including spacious homes equipped with handsome (and expensive) cabinet-style gramophone players.[26] In the United States, the market was also coded white. Operatic stages and audiences remained largely segregated throughout the short-playing era, the end of which coincided with the debut at the New York Metropolitan Opera in 1955 of Marian Anderson, the first Black singer to perform there. Black audiences supported pioneering Black institutions such as the Theodore Drury Grand Opera Company,[27] but in the recording industry, which defined both music and markets along racial and ethnic lines,[28] neither Black singers nor Black consumers figured significantly in the conception of opera. Anderson herself is a case in point. Though a prolific recording artist, she made few opera discs; instead, she sang art songs and spirituals, reflecting her own career as a concert performer but also distinguishing herself from her many contemporaries who recorded opera excerpts that they never sang onstage.[29] Records may have made opera more accessible, but not every community was invited to take advantage.

Women were not always welcome either, other than as beautiful voices emanating from the machine. Much as they do now, men dominated the industry, the journalism, and even the mythologies of listening: from Mann's Castorp to Nick Hornby's Rob (the record-collecting protagonist of the 1995 novel *High Fidelity*), the solitary obsessives are all male. Moreover, aspects of the technology put even the most venerated women performers at a disadvantage.[30] Vocal records are inevitably objectifying, splitting voices from bodies and smoothing the way for Castorp-like fantasies of possession. Itself coded masculine, the technology may even appear to "possess" the voice in its own right, as Wayne Koestenbaum suggests: "The phonograph, like psychoanalysis or the telephone or Cubism or Surrealism, was a male collaborative invention that played with the boundaries of voice and identity (often female or maternal), and made voice seem an emanation of male will."[31] For the *Gramophone* correspondent who preferred conducting at home over dancing, records may have been a way to filter social experience, to encounter femininity only when it was bereft of physical substance and safely under male control.

But the story does not end there. There have always been women connoisseurs of records, from the Miss Elsworth who presented *Rigoletto* to the Manchester gramophonists in 1924 (see above) to the women in Roy Shuker's 2010 ethnography of record collectors.[32] As Kenney shows, moreover, in the United States between 1890 and 1930, the consumer side of the industry included at least as many women as men, if not more. The campaign to make gramophones respectable was aimed especially at middle- and upper-class housewives, who were expected to manage culture in the home along with the household appliances; the gramophone was part of both. In the music retailers and department stores where records were sold, the clerks were predominantly women, and potentially equipped with the kind of specialist knowledge more typically associated with male collectors or gramophone operators.[33] If anyone actually used Edison's talking points to sell a record or suggested a companion disc for Marie Sundelius's "Batti batti, o bel Masetto," it may well have been a woman clerk talking to a woman customer, who later put the disc on the machine herself. Despite occupying their usual roles as engineers and public experts, men held no monopoly on record culture.

Production and Reproduction

While consumers may have known little about it, the recording process would also have shaped their listening experience. Most obviously, selections had to fit the lengths of the short-playing formats. Companies tried numerous alternatives, including both discs and cylinders that played longer than three to four and a half minutes, but it was the shorter durations that caught on.[34] Among operas, Mozart's adapted more easily than some; many numbers fit more or less comfortably within the usual durations or had natural stopping points if they needed to be split over two sides (e.g., the end of the opening allegro in "Madamina, il catalogo è questo"). Nonetheless, the constraint on length contributed to the paucity of complete recordings and to the relative avoidance of numbers that do not break down so neatly, such as Mozart's finales. It may also have affected decisions about tempo and timing. There is some evidence of artists singing faster on record than onstage,[35] perhaps to beat the time limits, although numbers could just as easily be cut (as in early recordings of "Il mio tesoro intanto"; see chap. 3).[36] More generally, performers may have adjusted pulse, phrasing, or pauses, whether taking less time or more, simply in order to make the best impression in an audio-only, vignette-length format.

Listeners also faced, and still do, uncertainties regarding recording and playback speed. Although 78 rpm was in theory the standard for most companies, in practice the number could vary in either direction, and players were equipped with speed adjusters.[37] Some efforts were made to advise consumers of this, but they were inconsistent and may or may not have been heeded. Playing a disc at a different speed from the one at which it was recorded affects both timing and pitch, and since the latter was itself not necessarily standardized (i.e., at A = 440 Hz), it can be impossible to determine what the correct playback speed of an individual performance should be. Engineers of modern transfers may take into account the practices of individual companies, the tuning conventions of different eras and locales, and other evidence. Their decisions, however, depend also on experience and taste, and they do not always agree. In the words of the engineer Roger Beardsley, "As you play the record, you have to ask yourself, 'does the voice sound like the voice you know?'—or, in the case of one you don't know, 'is it consistent with the vocal range and the music?' If it is, you can stick with it and be thankful; but if not, you can try moving a semitone on the appropriate side of the key. In the end, you must trust your instinct, in the knowledge that a certain amount of subjectivity is involved."[38] Early twentieth-century record buyers would have found themselves in a similar position, setting their players according to their own knowledge and preferences (and in some cases at least, intentionally trying out familiar pieces at different speeds).[39]

The effects can be experienced today with an adjustable turntable, sound-editing software, or—for better-known discs—a survey of competing transfers.[40] Pitched at A = 440 Hz, Antonio Scotti's 1909 recording of Don Giovanni's "Deh vieni alla finestra" sounds consistent, to my ears, with his other recordings as well as with contemporary descriptions of his style (see chap. 2). The delivery is crisp and clear, the vocal tone focused and forward, the vibrato moderate. A shift of 2.5 percent in either direction (roughly 2 rpm) leaves much of the character of the rendition intact. The changes in tempo are slight, approximately 1 bpm on either side of circa 50 bpm, and while the changes in pitch are more noticeable, about a quarter tone, the voice remains direct and energetic, even at the slower speed. Push another 2.5 percent in either direction, shifting pitch by a half step, and the mood alters substantially. Scotti is still recognizable as Scotti, but his tone is noticeably thinner, the energy edging toward strain—or just as noticeably thicker, richness edging toward sluggishness. The descriptions in part I should be read with this in mind. For those who played Scotti significantly faster or slower than I have (or who do so now), his Don Giovanni may have seemed either more insistent or more languid than the one I describe. ♫

The practices of acoustic recording, especially in its early years, had further ramifications for sound.[41] The physical vibrations produced by the musicians, focused by a recording horn, had to be strong and balanced, for the relative levels of the voices and instruments could not be adjusted later. The conventional arrangements of concert and stage performance did not produce the right results, forcing engineers to get creative. For vocal music, singers generally stood directly in front of the horn, moving backward on loud notes to avoid overpowering the machine's capacity for volume. Accompanists hovered behind and sometimes above: pianos, the sole instruments on many of the earliest opera recordings, were often perched on platforms behind the singers' heads, while ensembles sat on the floor or on risers of varying heights, arranged by volume so as to put the quieter instruments in front.[42] Even with that, conventional strings did not record well, so that basses were often replaced by tubas and upper strings by Stroh instruments, which had metal horns in place of wooden sound boxes. Accompaniments were rewritten, whether to suit the piano and smaller ensembles or to enrich textures that were perceived as overly thin (see chap. 2). Compared to concert or stage renditions, acoustic recordings also had a narrow dynamic range, and for all the ingenuity that went into managing volume and balance, the original technology only captured frequencies between about 150 and 2,000 Hz, or roughly a seventh below middle C to three octaves above. A lot of bass lines go missing or appear in the tenor register, and high notes lose many of their overtones, the upper harmonics that give each instrument and voice its unique timbre. Compression spelled a significant loss of color and resonance for women's voices especially.

Sound engineering improved steadily, and by the end of the acoustic era the need for special configurations and instrumentations had lessened. Still, the advent of electrical recording in 1925 brought profound changes. With sound no longer captured physically but converted into electrical impulses, it became possible to record a greater range of dynamics and, eventually, the entire spectrum of audible frequencies. Microphones could collect diffused sounds as well, allowing performers to work in more traditional configurations, and engineers could increasingly adjust the relative levels of different voices and instruments, both during and after recording. Not surprisingly, the years following 1925 witnessed a surge not only of new recordings but of rerecordings by singers who had released acoustic records and who presumably saw an opportunity to meet changing consumer expectations as well as to capture fuller images of their own vocal identities.

Yet despite the differences, the electrical process left key aspects of recorded opera unchanged. Discs continued to be manufactured from a mixture of shellac and carborundum, generating the surface hiss that distinguishes them from the quieter vinyl records of later years, to say nothing of noiseless CDs. Companies continued to sell discs primarily in plain cardboard sleeves, leaving listeners to find as much or as little information as they wished about the contents. And above all, electrically recorded discs still lasted only three to four and a half minutes. Even as the other great innovation of the 1920s, radio, made complete operas a more regular part of the media landscape, records remained quintessentially the province of the excerpt.

Their virtues may be easier to appreciate now than at any time since the industry switched to long-playing albums in the mid-1950s. Excerpt recordings have never entirely disappeared, living on in highlights releases as well as collections of favorite numbers performed by individual singers. More striking, the dominant formats of the last two decades have effectively reconstituted some of the conditions of the short-playing era. The vast majority of opera on the Internet circulates in excerpt form, and complete recordings can be downloaded or streamed piecemeal. Many digital files come with as little information as discs and cylinders did, and if context is now only a click away, it may be no more consistent, reliable, or related to the parent opera than when it came from record catalogues or sales clerks. Most remarkable of all, if they are compressed for download or streaming, digital files reintroduce a familiar limitation in audio quality. As anyone knows who has ever compared an MP3 track to its uncompressed original, some high frequencies go missing—not nearly so many as in the acoustic recording process, but enough to cause a palpable loss in resonance and color. Run the sound through a tinny set of earbuds or computer speakers, and the envelope shrinks still more.

That the fragmented, sonically variable result is a meaningful source of operatic experience helps put the short-playing era in perspective. *Werktreue* may have been a low priority for the early recording industry, which rearranged excerpts as it saw fit and devoted comparatively few resources to producing complete operas. Discs and cylinders may also sound low-fi to modern ears, so much so that another advertising ploy of the era, the public concert pairing live singers with their own recordings in an effort to demonstrate the gramophone's fidelity, seems optimistic at best.[43] Nonetheless, early records remain a technological and artistic marvel, an archive of style, emotion, and personality unsurpassed by any subsequent format. They preserve a wealth of performance history, including of *Don Giovanni*, and they typify the status of opera in the age of recording. To return to a metaphor from the introduction, short-playing records created the original operatic "cloud," a shifting aggregate of moments in which listeners could immerse themselves, with or without reference to parent works or to the sources and chronologies of the performances themselves. They could equally well imagine character portraits or songs, fragments of far-off dramas or expressions of intimate feeling unconnected to a story. Or both at once. Excerpts are opera in its most characteristically mediated form, simultaneously complete and incomplete.

CHAPTER TWO

Rhetorics of Seduction

Don Giovanni's "Deh vieni alla finestra" occupies a special place both in the opera and among *Don Giovanni* excerpt records. It shows the title character resorting to persuasion to get what he wants, as opposed to his more customary bullying or force. His sights fixed on Donna Elvira's maid, he sings a love lyric replete with flattery and romantic conceits. He "really" sings too, accompanied by mandolin and plucking strings suggestive of a guitar, a rare moment in which story and performance become one. As a recorded number, meanwhile, his tuneful and strophic canzonetta (as Mozart terms it) fits perfectly with the amorous ballads so beloved of early recording artists in general. This helps to explain its popularity: by 1955 "Deh vieni alla finestra" had been recorded more than eighty times and released together with sentimental favorites such as Martini's "Plaisir d'amour" (see chap. 1). One can imagine such records serving many scenes of persuasion, their sweet melodies and words imploring not the fictional maid so much as real-life consumers, or anyone else for whom the discs were played.

In the opera, of course, Don Giovanni is not merely persuading but seducing, using false promises and a disguise to lure the maid into sex. On its own the aria does not reveal the deception, which may have bolstered its appeal to an industry that fretted greatly about respectability. But whether they recognized it or not, consumers of "Deh vieni alla finestra" records bought into a regular primer on seduction, an encyclopedia of strategies with deep roots in the mythology of libertinism. Some performances exude warm sincerity; others, melancholic desperation; still others, the very forcefulness that Don Giovanni seems in this moment to have suppressed. The differences stem in part from the two great transitions documented by early twentieth-century opera recording: from acoustic to electric technology and from rhetorical to modernist performance. Yet difference stems equally from the imaginations of individual singers, most of whom played Don Giovanni onstage and brought some of their signature characterizations into the recording studio. Brevity and songfulness notwithstanding, "Deh vieni alla finestra" plumbs the theme of the opera like no other number. Its studied simplicity allows the rhetorics of seduction to multiply.

A Serenade Evolves

A pair of iconic discs captures the sweeping changes in opera performance that mark the short-playing era. One, made in 1904, features the French baritone Victor Maurel. Born in 1848, he belongs to the earliest generation captured on record, and his rendering suggests how Mozart may have sounded as early as the 1870s or 1880s, when Maurel was a leading Don Giovanni in the theater. The other disc, made in 1930, features the Italian bass Ezio Pinza, then near the beginning of a career that would make him the most famous Don Giovanni of the twentieth century. Born in 1892, he is separated from Maurel by two generations and nothing short of an artistic revolution. ♫

At first listen, it is the logistical differences that stand out most. The earlier recording is acoustic and the later electric, one consequence of which is that Maurel sounds more distant than Pinza. His intensity equals the later singer's, but the mechanical process and narrow frequency range muffle his voice somewhat, making it seem as if he were in another room. Microphones eased the compression problem and captured higher resonances, so Pinza's bass seems not only closer but also, paradoxically, brighter than Maurel's baritone. Equally striking is the difference between the accompaniments, furnished in one case by piano, in the other by the ensemble called for by Mozart, strings and mandolin. The contrast heightens the distance between the respective atmospheres, in the one case shadowy and claustrophobic, in the other shiny and spacious. Moreover, instruments and voices together point to differing standards of production. Rhythmically speaking, Maurel and his pianist are not always in sync, and they both make mistakes, notably at the beginning of the second verse, where the singer skips several beats and his partner has to scramble to catch up. By contrast, Pinza and his string ensemble are accurate and in time with one another throughout, reflecting a new level of polish in commercially released recordings.

Polish is a good measure for comparing the vocal performances as well. Flubs aside, Maurel's singing is no less impressive or engaged than Pinza's, but it is far less smooth. Every section of melody and even some notes and words have high points or low points or otherwise distinctive shapes, whereas Pinza's phrases are evenly contoured and consistent. Typical of rhetorical and modernist approaches generally, the difference is epitomized by a detail of the very first vocal entrance, an arching phrase that offers the possibility of adding an expressive ornament, a dissonant C♯, to a crucial downbeat: "Deh vieni alla fi-*ne*-stra, o mio tesoro" (Ah! come to the window, oh my treasure).[1] Maurel and his contemporaries were accustomed to improvising ornaments, and the baritone not only adds but extends and highlights the C♯, creating a tiny expressive drama. Pinza skips it, echoing

The Missing C♯

The opening phrase of "Deh vieni alla finestra" leaps from the fifth scale degree up to the tonic, A to D, then descends scalewise to the sixth degree, B. According to Mozart's autograph score, the B arrives without fanfare on the second downbeat, "fi-*ne*-stra," and so it is printed in most editions. In the next phrase, however, which begins with the same arch, the B is prefaced by a poignant dissonance, a C♯ suspended from the previous measure: "con-so-*lar*" (fig. 2.1). The discrepancy may be an oversight: in the second verse, which uses the same melody, Mozart writes a C♯ in both phrases, suggesting he either forgot to do so at the beginning or assumed that singers would insert the note on their own. Whatever its origins, on short-playing records the "missing" C♯ is a harbinger of style and generation. Victor Maurel not only adds it but leans on it, and rather than resolve to B on the second or third eighth note of the 6/8 time, as Mozart calls for later, he falls only at the fourth eighth note, a written rest. Aided by other details—a slight pause before the beginning of the phrase, a subsequent return to tempo, an audible swoop from A up to D, a shortening of the first syllable of "finestra" into a double-time pickup—after all this, the added ornament comes across as a rhetorical flourish, a glimpse of expressive bravura. By contrast, Ezio Pinza eschews both the C♯ itself and anything resembling Maurel's delays and anticipations and swoops: his rhythm and articulation remain steady and clean. The crucial syllable remains expressive—"fi-*ne*-stra"—but less as a note than as a function, a resting place on the downward side of the arch.

FIGURE 2.1 "Deh vieni alla finestra"
a. Opening phrase without C♯
b. Second phrase with C♯

his own contemporaries' rejection of ornament in favor of shape and structure. Here and in many subsequent recordings, the moment becomes a handy index of generational preference.

The C♯ is only the beginning. Ornamentation, by which I mean a whole range of additions and alterations to the written score, is one of several elements that separate nineteenth- from twentieth-century singers—or, more precisely, singers born before about 1885 from those born later. Among the soloists on Serenade records, nearly 80 percent of the earlier group add the unwritten C♯, compared to only 35 percent of the later. A similar evolution affects the end of the aria, where the last two phrases are separated by a written rest:

<table>
<tr><td>non esser, gioia mia, con me crudele,</td><td>my joy, do not be cruel to me,</td></tr>
<tr><td>[rest]</td><td>[rest]</td></tr>
<tr><td>lasciati almen veder, mio bell'amore.</td><td>at least let yourself be seen,
my beautiful love.[2]</td></tr>
</table>

With characteristic freedom, Maurel ignores the rest and connects "crudele" to "lasciati" with a swoop, creating another expressive drama (1:38–1:45). Eight of his contemporaries do something similar, and another fifteen leap to an unwritten high F♯ on "a-*mor*-e," replacing Mozart's low-key finish with a bravura cadence in the upper register. All of this disappears among singers born after 1890, as do other practices such as adding turn figures or unwritten dotted rhythms. Only one ornamental practice survives: sliding or swooping, more properly known as portamento, perhaps because it seems less like an addition to Mozart's text than a connective or technical device. Even Pinza incorporates an expressive slide near the end of the first verse ("tuoi," 0:43–0:48). Still, no one of his generation uses portamento as often or as obviously as Maurel, to say nothing of Mattia

Battistini (b. 1856), who transforms whole phrases into continuous curves. With time, "Deh vieni alla finestra" becomes more angular as well as less decorated.

And steadier. Pinza's opening phrases tick along evenly, while Maurel's alternately hold back and hurry forward. It is tempting to credit this to the earlier recording's use of piano, which would presumably have posed fewer problems of coordination than an ensemble, but there is more to it than that. Pinza sings generally a little faster than Maurel, just over 50 bpm to the dotted-quarter beat (i.e., the half measure of the 6/8 meter) as compared to the mid-40s. Overall speed, however, ranges widely even within generations, where singers contemporary with one another run the gamut from circa 35 bpm to circa 55 bpm. Tempo variation, on the other hand, follows a clear historical trajectory, one that becomes especially obvious in the third and fourth phrases of the aria, where the music and imagery grow more involved. The seducer's language switches from pleading to threats (or from flattery back to pleading, in the second verse)—

> se neghi a me di dar qualche ristoro,   if you refuse to give me some comfort,
> davanti agli occhi tuoi morir vogl'io.   I wish to die before your eyes.[3]

—and the music traverses the high point of the vocal line along with the number's most venturesome harmonies, complete with pungent accidentals. Figure 2.2 charts the two soloists' beat-by-beat progress through this passage of heightened intensity.[4] Maurel sets off more or less evenly but slows significantly on the high note, "ri-*sto*-ro," and even more so at the cadence, "vogl'-*i*-o," where the pianist waits obligingly. Pinza keeps the beat; though measurable, his anticipations on "qualche" and elsewhere are

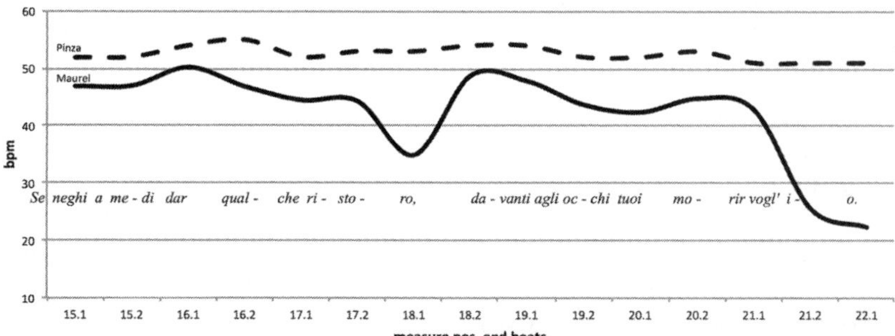

FIGURE 2.2 Tempo variation in "Deh vieni alla finestra," third and fourth phrases: Victor Maurel (b. 1848), Ezio Pinza (b. 1892)

modest and make little impression on the ear. He is not stiff, but his flexibility is on a micro scale whereas Maurel's is decidedly macro. The same is true of the respective accompanists. In addition to following the soloist's fluctuations, Maurel's pianist introduces still more, including an accelerando during the interlude between the first and second verses. Ending with an emphatic thump, it balances the ritardando of the preceding cadence and sets up Maurel to take his time beginning the second verse. Pinza's mandolinist plays the same passage with customary precision, neither speeding up nor slowing down and preparing the singer to resume right in tempo.

For purposes of comparison, figure 2.3 graphs the same two phrases in twenty additional recordings. Grouped by dates of birth, from 1856 to 1916, the soloists represent a cross-section of generations and also of national backgrounds—Italian, French, German, Austrian, Swedish, and Hungarian, whose representatives sometimes (though by no means always) share preferences. Mirroring the difference between Maurel and Pinza, their tempos become less variable over time. This is true both of the phrase endings, where the earlier generations tend to slow dramatically and the later less so or not at all, and of the beats within the phrases, where the use of rubato diminishes significantly. It is as if a prosodic ideal gives way to a poetic one, uneven and detail-rich sentences to patterned and mellifluous verse.

The sound ideal of performances changes as well, partly though not exclusively in concert with the shift from acoustic to electric recording. Before 1920 "Deh vieni alla finestra" was rarely recorded with its original accompaniment; performers used either piano reductions or arrangements for strings without mandolin. Both produce a heavier and more accented sound, even the ensemble scorings in which the mandolin part is typically played by pizzicato violins, which sound considerably more strident (especially if the recording uses Stroh violins, with their metal amplification

FIGURE 2.3 Tempo variation in twenty recordings of "Deh vieni alla finestra," third and fourth phrases

a. Mattia Battistini (b. 1856), Leopold Demuth (b. 1861), Jean-François Delmas (b. 1861), Maurice Renaud (b. 1861), Antonio Scotti (b. 1866)

b. John Forsell (b. 1868), Mario Sammarco (b. 1868), Giuseppe DeLuca (b. 1876), Titta Ruffo (b. 1877), Vanni-Marcoux (b. 1877)

c. Heinrich Schlusnus (b. 1888), Mariano Stabile (b. 1888), André Pernet (b. 1894), Willi Domgraf-Fassbaender (b. 1897), Paul Schöffler (b. 1897)

d. Karl Schmitt-Walter (b. 1900), Sándor (Alexander) Svéd (b. 1906), Tito Gobbi (b. 1913), Giuseppe Valdengo (b. 1914), Giuseppe Taddei (b. 1916)

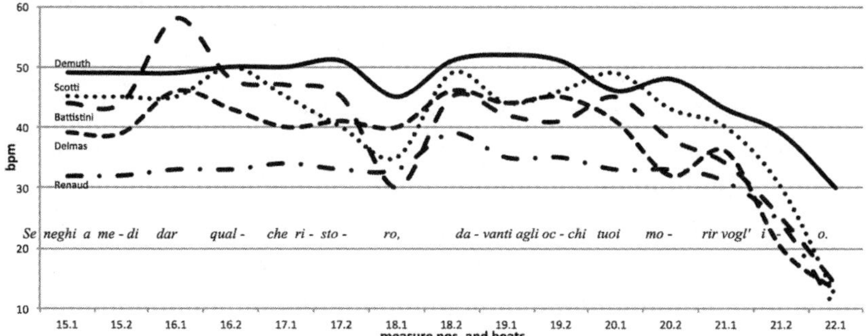
60
50
Demuth
Scotti
Battistini
Delmas
Renaud
bpm
40
30
Se neghi a me - di dar qual - che ri - sto - ro, da - vanti agli oc - chi tuoi mo - rir vogl' i - o.
20
10
15.1 15.2 16.1 16.2 17.1 17.2 18.1 18.2 19.1 19.2 20.1 20.2 21.1 21.2 22.1
measure nos. and beats

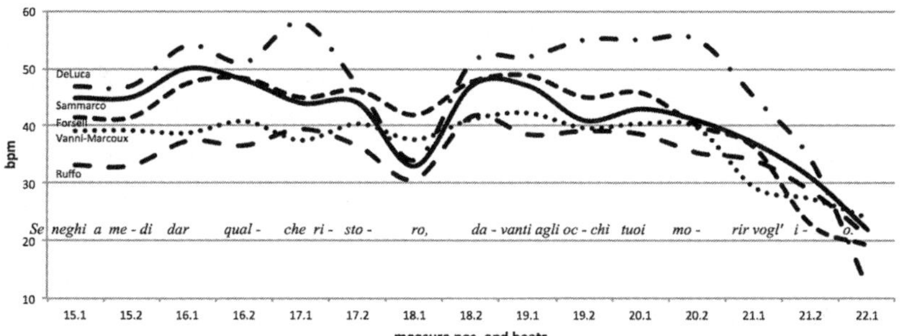
60
50
DeLuca
Sammarco
Forsell
Vanni-Marcoux
Ruffo
bpm
40
30
Se neghi a me - di dar qual - che ri - sto - ro, da - vanti agli oc - chi tuoi mo - rir vogl' i - o.
20
10
15.1 15.2 16.1 16.2 17.1 17.2 18.1 18.2 19.1 19.2 20.1 20.2 21.1 21.2 22.1
measure nos. and beats

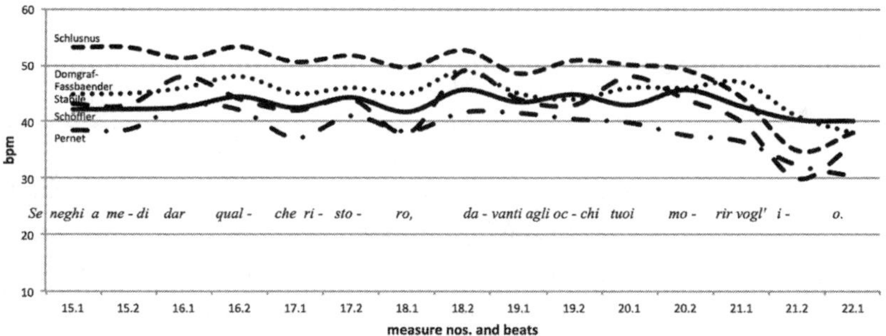
60
50
Schlusnus
Domgraf-
Fassbaender
Stabile
Schöffler
Pernet
bpm
40
30
Se neghi a me - di dar qual - che ri - sto - ro, da - vanti agli oc - chi tuoi mo - rir vogl' i - o.
20
10
15.1 15.2 16.1 16.2 17.1 17.2 18.1 18.2 19.1 19.2 20.1 20.2 21.1 21.2 22.1
measure nos. and beats

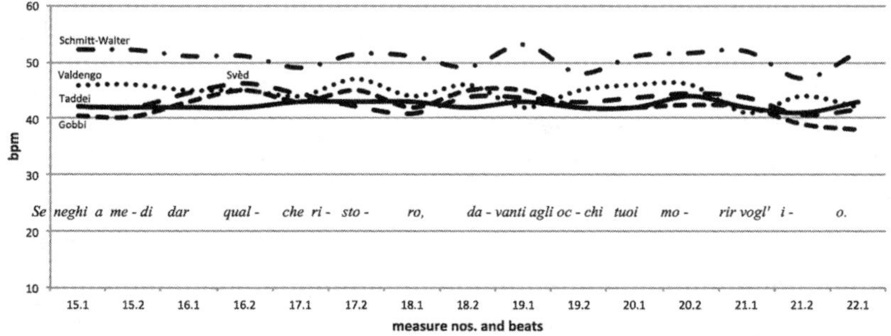
60
50
Schmitt-Walter
Valdengo Svèd
Taddei
Gobbi
bpm
40
30
Se neghi a me - di dar qual - che ri - sto - ro, da - vanti agli oc - chi tuoi mo - rir vogl' i - o.
20
10
15.1 15.2 16.1 16.2 17.1 17.2 18.1 18.2 19.1 19.2 20.1 20.2 21.1 21.2 22.1
measure nos. and beats

Tempo Variations

Recordings of "Deh vieni alla finestra" document both the general decline of tempo variation over the first half of the twentieth century and the individual nuances of particular soloists. At the ends of the verses, the singers of the earliest generations all take an enormous ritard, as Victor Maurel does (figs. 2.3a and b). Among Ezio Pinza's contemporaries, by contrast, the ritard diminishes and in some cases shifts from the second half of the penultimate measure to the first (fig. 2.3c), and among the latest generations the endings smooth out still more (fig. 2.3d). The aria's third phrase shows less consistent change. Singers born before 1885 tend to stretch "ristoro" by 10–25 percent, while their successors slow only slightly or not at all. Yet in this case, the consensus among earlier singers is not so strong: Leopold Demuth and Vanni-Marcoux both keep a comparatively regular beat through the entire third phrase (figs. 2.3a and b). Similarly, the use of rubato within the phrases diminishes over time, but not uniformly. The earlier generations include singers who push and pull at every beat, like Battistini and Jean-François Delmas, whereas later singers trend toward regularity. At the same time, contemporaries of Battistini such as Demuth, Vanni-Marcoux, and even Maurel, though they are not so metronomic as Pinza or Giuseppe Taddei, nevertheless maintain a relatively even pulse midphrase, saving variation for the beginnings and ends.

horns). Many ensemble arrangements also reinforce or replace the strings with brass or woodwind instruments, including tubas on the bass line. The aim was to generate sufficient oomph for the acoustic recording process, but several recordings add more than was necessary for that, notably woodwind parts that double the violins or support their harmonies with long tones. To ears accustomed to Verdi and Wagner, or to the lush arrangements of popular ballads, Mozart's little string band must have seemed to lack fullness.

The original scoring gained traction during the early 1920s and became the standard with the advent of electrical recording, which captured the delicate tones of the mandolin, in particular, better than the acoustic process could. Again, however, the transformation goes beyond what the change in technology can account for. Compared to the thumping pianos and splatting tubas of old, the strings trip along lightly, and some recordings capture them in a dry sonic envelope that foregrounds their transparency. Pinza's version is a good example, clear and bright. Other recordings, however, introduce reverberation, which transports the performers into a more or less echoey space and also takes the edge off the plucking string attacks, including those of the mandolin. Still others eliminate the mandolin entirely, substituting a less crystalline guitar or harp. In Paul Schöffler's recording of 1947, warm reverb and a harp dissolve the whole performance into a shimmering haze, as if the ideal were no longer Mozart or even Verdi and Wagner, but rather Debussy.

Vocal delivery undergoes a related evolution. Voice type tends to be distinguished within generations rather than across them: at any given time, "Deh vieni alla finestra" was recorded by a variety of baritones and basses. Delivery, on the other hand—the manner of singing as opposed to the voice itself—develops audibly over time, becoming progressively lighter alongside the increasingly transparent accompaniments. Not surprisingly, perhaps, soloists on acoustic recordings tend to sing heartily, like Maurel. This instinct does not disappear in the electric era; Pinza, for one, still sings as if he wants to reach the back row. Already in acoustic recordings, however, one can hear quieter approaches, achieved with less emphatic attacks, more tapered phrase endings, and the use of head voice. Some soloists may even have regarded tempo variability as a means of mitigating otherwise forceful dynamics and tone. In the electric era a more restrained manner comes to predominate, and eventually the pendulum swings once again beyond what the technology in itself would seem to justify. Perhaps taking their cue from popular crooners like Rudy Vallee and Bing Crosby, soloists such as Roger Bourdin, Karl Schmitt-Walter, and Giuseppe Taddei use microphones to create the illusion of romantic intimacy.[5] They sing at a level that would hardly carry in a theater, murmuring their persuasions in real rather than stage whispers. The Serenade moves from the stage, or the imagined street below the beloved's window, right into the bedroom itself.

Images of Seduction

I do not want to read too much into generational changes; soloists of a given cohort may sound as different from one another as they do from their predecessors or followers. Nonetheless, the distance traveled between Maurel (b. 1848) and Taddei (b. 1916) does have suggestive implications. As Robert Philip and others have shown, the entirety of opera and classical music performance underwent an epochal transformation in the latter decades of the nineteenth century and the first half of the twentieth.[6]

Whether they were singers or instrumentalists, contemporaries of Maurel tended to delineate musical and emotional details with flexible rhythm and ornamentation, among other means, while those of Pinza came to focus more on articulating phrases and forms, in tempo and without decoration. Following the protomodernist critic T. E. Hulme, Richard Taruskin labeled the two tendencies "vitalist" and "geometric," the one associated with Romanticism and the fluctuating pulsations of the inner life, the other with modernism and the regular patterning of its favorite non-Romantic artforms, from fixed-form medieval poetry to skyscrapers.[7] Nicholas Cook adopts the terms "rhetorical" and "structuralist," connecting the first to "the communication of moment-to-moment expressiveness" and the second to "the reproduction of an ideal, essentially atemporal subject"—that is, the "work" in its original gestalt.[8] Needless to say, the categories are not mutually exclusive, nor do they follow one other like chapters in a book. They are the opposite poles of an uneven, incomplete, yet still observable reorientation from emotion toward form, from details toward the whole.

The consequences for "Deh vieni alla finestra" are in some ways obvious. As singers sing more evenly and with fewer vivid effects, and instrumentalists exchange accents for shimmer, we can easily imagine the serenader becoming a cooler customer. He loses—or sublimates, maybe—whatever Romantic energy animates the ups and downs of the earliest records, substituting a modernist urbanity that absorbs even the heightened intensity of the third and fourth phrases into a smooth stream of melody. He may also become more of a "seducer" in the traditional sense of the term. Don Giovanni has not always been understood as such, even though his abandoned spouse, Donna Elvira, credits him with a lengthy campaign of "craftiness, oaths, and flattery" preceding their sham marriage.[9] Her description recalls any number of libertine texts, from the first-century *Art of Love* by Ovid to the 1843 "Seducer's Diary" of Søren Kierkegaard, which appeared in the same volume as the philosopher's famous essay on *Don Giovanni* and chronicles a months-long process of surveillance, notes, conversations, quarrels, reconciliations, and more.[10] In the opera, by contrast, Don Giovanni shows little taste for process, wooing Zerlina, Donna Elvira's maid, and Donna Elvira herself (in the act 2 trio, "Ah taci ingiusto core") quickly and without apparent calculation. Rather than psychology, he wields force of personality—and when that fails, physical force. Kierkegaard distinguishes such directness from his diarist's more careful manipulations: "[Don Giovanni] desires, and this desire acts seductively. . . . He lacks the time to be a seducer."[11]

Kierkegaard's assessment has had long echoes in both performance and criticism, yet it is not necessarily the last word. Metaphorically at least, "structuralist" performances of "Deh vieni alla finestra" stand on the side of intellect, and the calculated projection of phrases, such as can be heard on many later recordings, may well suggest a carefully planned deception. In its classic formulation, modernism also presumes a degree of emotional detachment often associated with seduction. In a passage quoted by Taruskin, the archmodernist Stravinsky (himself following Pierre Souvtchinsky) declares that "music thus gains strength in the measure that it does not succumb to the seductions of variety. What it loses in questionable riches it gains in true solidity."[12] Stravinsky's "music" sounds a lot like Kierkegaard's diarist, maintaining a discipline that guards against emotional entanglement. Where performance is concerned, rhythmic or tonal consistency may well suggest a similar disengagement.

On the other hand, if the structuralist Don Giovanni seems craftier than his rhetorical predecessors, the transformation is neither one-way nor absolute. As noted above, plenty of rhetorical performances incorporate steady rhythms and clean lines, which may suggest ample measures of self-possession.[13] More fundamentally, the modernist equation of consistency with emotional restraint does not always hold up, especially where opera is concerned. Even Stravinsky credited his preferred form of music, the kind without "variety," with inspiring "a feeling of euphoria and, so to speak, of 'dynamic calm.'"[14] Cool-headed as those emotions may be, they are still emotions, and in an opera about sex and power, even cool performances necessarily stir up the passions. After all, every rendering of "Deh vieni alla finestra" must be rhetorical in the original sense of the term, which is to say, persuasive: it must still endeavor to convince the maid to come down from the window, or the consumer to buy the record. Modernist singers do not eject feeling so much as channel it into new stratagems, just as their predecessors put emotional nuance in the service of their own varying tactics. Wherever it falls historically, a recording's impact depends equally on its idiosyncrasies.

The Rhetorical Don

The *Gramophone* critic P. G. Hurst would have agreed. In 1935 he devoted part of his "Collector's Corner" to old recordings of "Deh vieni alla finestra," including Maurel's along with those of four singers born in the 1860s: Mario Ancona, Maurice Renaud, Antonio Scotti, and Mario Sammarco. His survey demonstrates how quickly records created a historical consciousness about performance, and also how differently singers of the same era could come off, even to an aficionado. Hurst discusses the records in terms of the parent opera, first sketching the soloist's stage interpretation of Don Giovanni, then examining their performance of the aria as "an interesting commentary on the methods of the great baritones in presenting this most difficult character."[15] What he hears is not a single character but several, a kind of rogue's gallery of serenaders. ♩

Dotted Rhythms and High Notes

Nothing separates "rhetorical" serenaders from one another more than dotted rhythms and high notes, whether written or improvised. The highest written pitch of "Deh vieni alla finestra," E, falls in the middle of a three-note dotted figure: long–*short*–long. Singers at the forceful end of the spectrum, such as Maurel or Sammarco, mark the arrival by slowing down and savoring all three notes: "ri-*sto-o-o*-ro" (Maurel: 0:38–0:48; Sammarco: 0:44–0:57). By contrast, the smoother Maurice Renaud prolongs the first note (D) and barely touches the second (E), keeping the moment as smooth as everything else in his performance. Antonio Scotti takes yet another tack, rendering the dotted figure on "ristoro" crisply and introducing other dotted rhythms where Mozart does not call for them. Not once but several times, he delays the last note of one measure to create a double-time upbeat to the next, strengthening his projection of lightness and restraint ("Deh vieni alla . . . *fi*-ne-stra"). Similarly, at the final cadence, rather than extend the written high D (as Maurel does) or vault up to high F♯ (as Sammarco does), Scotti sustains F♯ in the lower register and descends quietly to D. The sensuous Battistini, finally, goes in the opposite direction and removes some dotted figures; in his 1902 recording, the second syllable of "ri-*sto*-ro" flows across three even eighth notes in one long slide (0:35–0:45). Accordingly, his final cadences forego power in favor of mystery, even when they incorporate an added high note. In both 1902 and 1913 he ignores the rest separating the last two phrases of the aria, but rather than swoop up and begin the final statement with a punch, as Maurel does ("crudele—*la*-scia-ti"), he pulls back. The energy builds over the next few beats but dissipates again at the cadence ("a-*mor*-e"), where in 1913 he descends into the lower register as written, and in 1902 leaps to high F♯ delivered sotto voce, a startling choice that undercuts any bravura effect (1:38–2:07).

He begins with Sammarco (b. 1868), whom he admits to having never seen as Don Giovanni, but whom he remembers as being "blunt in his methods, almost brutal, and [making] his characters appear short-tempered and intolerant." On record, "his rendering of the Don's Serenade supports this view, and gives the impress that, having made this concession to custom, he would expect no coyness or hesitation on the part of the lady. The high note which ends it gives it almost the air of a challenge." Further on Maurel elicits a related if more positively worded description: "a Don in the full pride of his prime, of dominating personality, supremely confident in himself. He was a patrician to his finger-tips, and quite overpowering in the splendour of his appeal."[16] Recalling the discussion above, the two singers seem to have reminded Hurst of something like Kierkegaard's vision of the character, an avatar of desire with no patience for seduction. Sammarco's recordings of "Deh vieni alla finestra," of which he made several, fit the description: reminiscent of Maurel in his uneven tempo and attention to detail, he brings a brighter and edgier tone to the aria, with more open vowels and a prominent vibrato, and to the older singer's rhetorical ploys he adds aspirations, little gulps for air that set off key attacks. As Hurst says, he also crowns his performances with an exceptionally aggressive cadence, belting out an unwritten high F♯ before cutting off abruptly on D. Emotional power shades into force.

As for Maurel, his own views of the character were not unlike Kierkegaard's. Borrowing the terms of evolutionary theory, he argued in 1897 that the "natural mission" of Don Giovanni is to be a "procreator," "one of the beautiful animals destined to perpetuate the species."[17] He is not an animal, per se, in the sense of acting only by brute force (this was the view of Maurel's contemporary Alexandre Dumas *fils*, whom he goes some lengths to refute[18]), but what drives him is biological rather than intellectual, an instinct for life. The singer's performance does not entirely measure up; known as a "singing actor," Maurel articulated his roles with carefully planned gestures, vocal as well as physical, and though his Serenade does not break into speech (one of his signature techniques), its ornaments, accents, and tempo changes do suggest a thoughtful attention to detail—in short, a cannier intelligence than his assessment of Don Giovanni would seem to allow.[19] On the other hand, the accumulation of detail may have a different impact in a love song than in the dramatic tours de force for which Maurel was most famous, such as Iago's monologue (from Verdi's *Otello*). For all that they differentiate words and phrases, neither Maurel nor his pianist really disrupts the prevailing lyricism of "Deh vieni alla finestra": the melodic arches and pulsating accompaniment bend without breaking, and the rhetorical effects seem like flashes of emotion trying to burst through the form. Sign of age though it may be, the audible straining of Maurel's voice reinforces the emotional intensity, as does rising agitation in the second verse, which culminates in that swooping connection of the last two phrases (see above). Whatever his inclinations elsewhere, as a serenader Maurel leans toward fervent lover, unreflective and demanding.

Hurst took nearly the opposite impression from a recording by the French baritone Maurice Renaud (b. 1861). Forcefulness gone, Don Giovanni now affects "an intense, almost pleading note; he seems actually to be in love, and to be pouring his whole soul into the business."[20] Renaud's tempo lags Maurel's by several points, and the relaxed speed goes along with mellifluous phrasing and a lush accompaniment; this is one of the arrangements that adds sustained chords for woodwinds. When he does aim for emphasis, moreover, he does so by gasping for air, as if the depth of his feeling were making him sob. No wonder Hurst detected a fresh note of conviction, as did witnesses to Renaud's live Don Giovanni: "Vicious as is his passion," wrote Abbie H. C. Finck, "it is sincere, and a world of feeling, almost of love, wells up in his eyes and voice."[21] At the same time, the key term here is "almost," and Hurst goes on to read sincerity as a strategy in itself: "Herein lies the guile of this trained and expert libertine, who could appreciate the satisfaction of a fish skilfully landed, or a game well won. He was a connoisseur in the gentle art; he was precise in his methods and left nothing to chance. So thorough was he, in fact, that he added a third stanza to his Serenade!"[22] The final witticism does double duty, undercutting the pretense to sincerity and reminding readers that the Don Giovanni in question is not merely fictional but a product of the recording process. On disc, "Deh vieni alla finestra" was brief enough to leave room for additional music, and several releases include fragments of recitative, extended instrumental preludes, or other arias. What Renaud adds is not in fact a third verse but a repetition of the second, sung now in the original Italian rather than the French translation used in the first two-thirds of the disc. The artifice involved in making a record, and in singing a "truthful" serenade, becomes all the more obvious.

According to Hurst, his next example, Antonio Scotti (b. 1866), played yet another version of Don Giovanni: "A fine, upstanding figure, of perfect breeding and with an easy good humour. He could show that sort of familiarity towards his inferiors which would never invite too ready a response: a man of taste and culture, and a magnificent companion . . . his Victor recording of 1905 is a perfect piece of work, and exactly suited to that well-groomed and fastidious quality in his voice."[23] Once again Hurst echoes the reports of theatergoers. Scotti was the Metropolitan Opera's Don Giovanni from 1899 until 1908, and American critics repeatedly praised his elegant bearing and meticulous singing. "He possesses a beautiful baritone voice," wrote Henry Krehbiel, "fine and smooth in quality, fluent in execution and managed with admirable skill. . . . He sings with intelligence and discrimination."[24] His recordings of "Deh vieni alla finestra" confirm the descriptions insofar as they project considerable finesse. Scotti sings without strain, his voice bright and thin like Sammarco's but without the edge. His tempos are brisk, around 50 bpm, and his expressive effects subtle: he avoids gasping, uses slides and accents sparingly, and extends but does not alter the cadences. Still very much in the rhetorical tradition, Scotti nevertheless carves out a distinctive identity by projecting more cultivation than intensity.

Not every recording of the number suggests such disparate conceptions as these. Hurst has little to say about his final example, Mario Ancona, perhaps because his strong tone and forthright delivery recall the fervency of Maurel and Sammarco. So do the performances of singers not discussed by Hurst, including Maurel's contemporary Jean Lassalle (b. 1847) and the younger John Forsell (b. 1868). Along similar lines, Jean-François Delmas (b. 1861) shares in Renaud's gasping sentimentality, while Leopold Demuth (b. 1861) and Charles Gilibert (b. 1866) echo Scotti's comparative lightness and restraint. Every singer adds personal nuances, but many of the overall characterizations approach one or another of Hurst's descriptions. Among singers trained in the late nineteenth century, the Serenade tends to elicit bluster, tears, or elegant reserve.

Unless, that is, the singer is Battistini, whose absence from Hurst's survey is surprising. Though less strongly identified with the role than Maurel or Scotti, he was arguably the most famous baritone of the era and a frequent Don Giovanni onstage. He also made multiple recordings of "Deh vieni alla finestra," and the versions from 1902 and 1913 stand out for their distinctly hypnotic eroticism. Battistini sings lightly and sometimes ethereally, anticipating the crooning of singers in the microphone era. He swells but rarely accents, uses more portamento than anyone else, and varies the rhythm constantly, pressing and releasing not just at cadences but within phrases, especially in the second verse, where the flattery in the text seems to inspire him to quasi-improvisatory freedom. Aided by accompaniments that follow his every nuance (and in 1902, by a pianist who lingers over the prelude and interlude rather than racing through like so many others), his performances have a silky smoothness that goes well beyond Scotti's restraint. Battistini is supposed to have been told that his Serenade constituted an "offence against public morals," presumably a reference to its latent eroticism.[25] His interpretation persuades by mesmerizing, casting the serenader in the role of the original tempter—the snake.

The Structuralist Don

At the other end of the short-playing era, singers spun new images, sometimes referencing the bullies and enchanters of old but within a transfigured musical context. By the time of his last recording of "Deh vieni alla finestra," made in 1924, even Battistini—now almost seventy—sounds like a different man. The portamenti and rhythmic quirks remain, but the tempo has steadied and the delivery grown more blunt. Five years later Pinza inaugurated his own forthright interpretation with a performance at the Met. Critics had their doubts, as Oscar Thompson's reaction attests: "Ezio Pinza's portrayal of the titular role suggested a roystering guardsman or a captain of the King's navy more than it did the loverly hidalgo of the Mozartean tradition, notwithstanding his costumic opulence. He had the inches but not the gait, the beard but not the manner. He made love by a variety of rude force, even though he contrived to sing 'La Ci Darem' [*sic*] and the Serenade gently."[26] Stark Young declared similarly that "Mr. Pinza, in a sort of public way, did very well with much of Don Juan. . . . But Mozart would have mourned the shortcoming of this Don Juan in elegance and aristocratic clarity and pride."[27] For New Yorkers, the standard of comparison may still have been the more elegant Scotti; though he had not performed the role since 1908, he remained in the Met company, and by Pinza's account the earlier production—conducted by Gustav Mahler, no less—had not been forgotten (he was the first singer to assume the role after Scotti).[28] Or perhaps critics were thinking of singers like Titta Ruffo or Vanni-Marcoux (both b. 1877), lighter-voiced baritones from the intervening generations whose recordings maintain some of the old traditions of rubato and ornamentation. Pinza's deep bass and no-frills delivery came as a shock.

He himself spoke of refining his conception during his long collaboration with the conductor Bruno Walter, which began in 1934.[29] To judge from two radio broadcasts conducted by Walter, however, dating from 1935 and 1942, Pinza never became a "loverly hidalgo." Walter (b. 1876) was a neoclassicist when it came to Mozart. Recalling a Viennese *Don Giovanni* of 1905, also conducted by Mahler and featuring a revolutionary set by Alfred Roller, he objected to the "luxurious red" of a bed of roses: "I realized that

Mozart's temperate orchestral coloring was ill suited to so rich a hue and that, generally speaking, the scenic picture of a Mozart opera ought to keep at a respectful distance from the work itself. A certain reserve and modesty seemed indicated by Mozart's dramatic-musical style."[30] His *Don Giovanni* is accordingly steady and precise, its emotional pitch governed more by tempo than by tempo variation or rhetorical effects. If Walter influenced Pinza, he did so in a way that accommodated the regular beat, even tone, and fidelity to the score of the singer's 1930 recording.

That same recording is among the fastest of the pre-1955 versions, which raises some unexpected connotations. Together with the unornamented consistency, the speed turns attention from the measure-by-measure details toward the phrases and the overall form, and it also conveys an urgency compounded by Pinza's noticeably fast vibrato. Steadiness notwithstanding, Pinza's serenader appears gripped by nervous energy, barely able to contain his impatience. In this he bears an affinity to Maurel's and Sammarco's characters, however differently they go about making their case. All three seek to overwhelm, and indeed Hurst's description of Sammarco's Don Giovanni—"short-tempered and intolerant . . . expect[ing] no coyness or hesitation"—may equally well fit Pinza's. Toward the end of the singer's career, Olin Downes described his characterization as harking back to "the adventurous, defiant, picaresque figure of the old legends which long preceded the opera."[31] More than entice or enchant, this Serenade presses forward with thinly veiled aggression.

It had few parallels in its own time, at least on record. Slightly older singers, such as Luigi Montesanto (b. 1887) and Mariano Stabile (b. 1888), achieve a similar forcefulness with deep voices, quick tempos, and uncluttered deliveries; they are among the first to refrain from adding the C♯ ornament to the first phrase. But generous rubato and a flair for detail put them stylistically closer to Maurel and Sammarco, or at least to the decade-older Ruffo, than to Pinza. Many other singers born after 1885 take another direction entirely, toward the quiet facilitated by electrical recording. A 1926 version by Heinrich Schlusnus (b. 1888) is a good example. Against a dry and transparent accompaniment, he sings at medium strength, not yet crooning but no longer aiming for the balcony. His delivery follows a middle path between rhetoric and structure: he still sings the C♯ and emphasizes with accents, but the majority of his articulations sound mild rather than forceful. Even at the cadences, a substantial ritard makes way not for a bravura roar, but for a relatively tranquil ending in the lower register. By 1935, in another recording of the same number, the ritards all but disappear and the delivery becomes even quieter, especially in the second verse, where he verges on sotto voce, a familiar effect in post-1950 performances. More variegated than Pinza's reading and in that sense more dramatic, Schlusnus's leave a much gentler impression, with little hint of urgency or force. ♩

Much the same can be said for other relatively calm interpretations. The singers range across the map, from German contemporaries of Schlusnus such as Heinrich Rehkemper (b. 1894) and Gerhard Hüsch (b. 1901), to the noted French Don Giovanni André Pernet (b. 1894), to the Italians Tito Gobbi (b. 1913) and Giuseppe Valdengo (b. 1914). Individual differences aside, they portray a serenader not dissimilar to Antonio Scotti's, restrained and subtle. Recordings with pronounced reverb, or with harp or guitar instead of mandolin, conjure a further variation on the scene. Mozart's original scoring isolates the voice by way of difference—sustained melody versus plucking backdrop. Reverb sustains everything, at least a little, and swapping the mandolin for instruments with a longer decay has a similar effect. The conceit that the singer is accompanying himself recedes, making way for a more impressionist use of sound. In Schöffler's (b. 1897) recording or in Roger Bourdin's (b. 1900), both with harp, the background combines with restrained vocal performances to create a new sensuousness, a velvety warmth that serves as a means of persuasion in its own right. Listeners are being hypnotized again, much as in Battistini except that now the whole ensemble is involved, not just the voice.

Taddei (b. 1916) also sings with restraint, but the use of the original accompaniment on his recording, played with exemplary precision, points toward one final image of the serenader. Anticipated in some respects by other singers, notably Karl Schmitt-Walter (b. 1900), Taddei aims at intimacy. His tone is light, almost falsetto, the tempo relaxed, and the means of expression subtle: in the second half of each verse, the site of so much emoting over the years, Taddei offers only the most modest of swells and diminuendos. Eschewing histrionics, threats, and magical obfuscation alike, his performance affects a kind of Mozartian pillow talk, the quiet reassurances of a lover who has already gained the listener's trust. It recalls a different image of Don Juan, this one from Henry Roujon's *Miremonde* (1887), a text quoted approvingly by Maurel despite his own quite different interpretation.[32] This libertine's ruse is to pretend friendship, as with Donna Elvira: "Our encounter, begun with high voices in a tone of decent coquetry, finished in complete quiet with a murmur of confidence, full of confessions, silences, and sighs."[33] So clever is the deception that the victim does not know she has been deceived.

Or does she? Recorded in 1955, the tail end of the short-playing era, Taddei's disc invites us to reflect on the dual roles of "Deh vieni alla finestra" as operatic outtake and independent love song. Heard as the former, with the dramatic situation firmly in mind, the innumerable versions of the aria seem like one vast catalogue of false promises. Surveying them is akin to reading Kierkegaard's "Seducer's Diary," or better yet Ovid's *Art of Love*, with its tips on the uses of eloquence, humility, manliness, elusiveness, flattery, pleading, weeping, friendliness, persistence, and—with disturbing casualness—violence.[34] Ovid does not imagine that all of these will figure into a single seduction; his text is a catalogue of libertine strategies. Serenade records are its musical equivalent. From rhetoric to structure, fervent straining to friendly whispering, they outline every trick a singing seducer might need.

Yet the same records, heard without reference to the opera, recall another scene of persuasion I mentioned at the beginning: that between singer and consumer. Their relationship is more coequal than Don Giovanni's with the maid, if still problematic: thrown together by technology and the market, they are both caught up in the commodification of sentiment and the alienation of performers from listeners. Nonetheless, listeners have an agency that a libertine's operatic victims do not, at least when the listeners have some control over what comes out of the gramophone. Taddei's quiet, ingratiating disc seems uniquely suited to this circumstance, offering a specifically "gramophonic" serenade. During the same year of 1955, the baritone sang the number again on one of the earliest LP recordings of the complete *Don Giovanni*, and the difference is striking.[35] Though his voice has the familiar half-whispered restraint, a quicker tempo and bouncier accompaniment (fronted by a very twangy mandolin) short-circuit any projection of intimacy. The serenader seems quiet but not confiding, outside rather than inside, dramatic more than domestic. Intentionally or not, the excerpt version turns a public moment into a private one and a performative plea into a friendly ask, almost a chat. It deceives only insofar as its emotional warmth, the feel-good romantic current that runs through so many records from the early twentieth century, emanates from a machine and not a person. Then again, unlike libertines, machines stay put. As Hurst sensed as early as 1935, the persuasions of Serenade recordings are artificial but permanent.

CHAPTER THREE

Demons and Dandies

When the Salzburg Festival made its 1954 production of *Don Giovanni* into a film, it immortalized a symbolic contrast in male dress.[1] The title character, played by Cesare Siepi, sports solid-colored doublets with minimal decoration, open at the neck and short enough to expose powerful legs in tights and boots. Don Ottavio, played by Anton Dermota, wears a brocade doublet and cape festooned with ruffles and an enormous wing collar, set off by knee pants and buckled shoes (fig. 3.1). Both costumes signal "Renaissance aristocrat," but they mark their wearers as different kinds of men: the libertine looks active, bold, and virile, while his opposite number is constricted, reserved, and proper. The dichotomy is borne out by the stage direction, which has Siepi moving vigorously in nearly every scene and Dermota remaining composed and often stationary.

Whether it meant to or not, the production realized one of the most venerable commonplaces of *Don Giovanni* criticism: that the title character represents manly initiative while Don Ottavio embodies effeminate passivity. The stereotypes can be heard as well as seen, including in the half century of excerpt recordings that precede the film. Performances of what critics considered Don Giovanni's signature number, "Fin ch'han dal vino," often teem with libidinous drive, while those of Don Ottavio's, "Il mio tesoro intanto," luxuriate in romantic lyricism. Yet the records also pack some surprises, singers who make the one aria more playful than aggressive, and the other more forceful than devoted. Whatever critics said, the received wisdom about the Dons was still up for debate in the early years of recording, suggesting that the commentaries in which it originated need to be reread,

FIGURE 3.1 Images of masculinity in *Don Giovanni*, dir. Graf 1954. DVD Deutsche Grammophon 0730199, 1994/2001.

a. Don Giovanni (Cesare Siepi)

b. Don Ottavio (Anton Dermota)

and its persistence into more recent productions questioned. In the quintessential opera of male conquest, masculine identity is anything but stable.

Devilry, Biology, Comedy

By the time of its earliest recordings, "Fin ch'han dal vino" had served for many years as an unlikely subject for character analysis—unlikely, because it fits no one's idea of self-revelation. Nicknamed the Champagne Aria for a stage tradition of raising (and often tossing) a glass, the number sketches a blunt narrative of drinking, dancing, and sex, all of it dedicated to aggrandizing the libertine's ego: "Ah la mia lista doman mattina d'una decina devi aumentar" (Ah, tomorrow morning you will have to increase my list by about ten).[2] He betrays no conflicted emotions or hidden motives, none of the inner complexities probed by so many of Mozart's other arias. Elsewhere, too, he talks about himself only in ironic recitatives and sings in whatever tone he needs in order to be persuasive: pastoral and down-to-earth for the lower classes (as in "Deh vieni alla finestra"), serious and high-flown for the aristocrats. By Mozartian standards he has no musical identity at all, no style to pin down his personality or social standing.

None of this prevented nineteenth- and early twentieth-century commentators from musing on his true nature, nor from reading their conclusions into "Fin ch'han dal vino." On the contrary, the character's very lack of definition seems to have been an inspiration. The speculating begins in E. T. A. Hoffmann's story of 1813, "Don Juan: A Fabulous Event That Happened to a Traveling Enthusiast," a text so influential that most later writers felt obliged to reckon with it, even if they disagreed with its claims. An impressionable narrator, ostensibly Hoffmann himself, comes away from a performance of the opera believing that he has witnessed an elevated morality play, a "conflict between divine and demonic forces."[3] During Mozart's overture, he says, "I saw fiery demons reaching out from darkest night with their glowing claws" (17), and when the protagonist/soloist appears (Hoffmann blurs the two), his magnificent appearance conceals further devilry: "Don Juan unwinds his cloak and stands splendidly dressed in red cut velvet with silver embroidery. A strong, noble figure, his face is beautifully masculine, with a prominent nose, piercing eyes, and softly formed lips. The peculiar twitch of a muscle over the eyebrows momentarily brings something of Mephistopheles into his physiognomy, which, without robbing the face of its beauty, triggers an involuntary shudder. It is as if he can practice the magic art of the rattlesnake; it is as if women, once gazed upon by him, could no longer leave him and—having been seized by his uncanny power—must complete their own destruction" (18). Though it is most notorious for including Donna Anna among Don Giovanni's conquests (see chap. 9), the subsequent account makes equally bold assertions about the libertine himself, who, despite the initial description, turns out to be less an embodiment of evil than its agent—and a troubled one. Driven by an "eternal, burning desire" to pursue "all the experiences of the earthly life," he is tricked by the devil into thinking he can satisfy his urges by seducing women (25). When he cannot, his frustration bursts forth in "Fin ch'han dal vino," where, in Hoffmann's description, "Don Juan quite blatantly expressed his inner, tortured self, and his mockery of the little people around him, who were put there only to satisfy his desire to interfere in their dull actions and doings. His forehead muscle twitched more powerfully than before" (19). Far from lacking depth, the aria exposes the character for who he really is, a vessel of Mephistophelean wickedness who copes by victimizing others. Facial tic and "wild" music strip the cover off a tormented soul, making the libertine's rhapsody as psychologically revealing as any conventional soliloquy.

The demonic Don Giovanni would go on to have a long literary career.

He appears again in Søren Kierkegaard's *Either/Or* of 1843, now folded into a broader philosophy of male eros. Kierkegaard's hero is not conflicted—in that respect, Hoffmann's closest heir is the Freudian Otto Rank, whose *Don Juan Legend* of 1924 diagnoses the libertine with an Oedipus complex. Kierkegaard, by contrast, regards desire and sensuousness as transcending any individual history, to say nothing of authentic human connection or reciprocity. Don Giovanni and his medium, music, are conduits of an impersonal and implicitly violent eroticism: "Desire is absolutely genuine, victorious, triumphant, irresistible, and demonic. . . . The issue here is not desire in a particular individual but desire as a principle. . . . This is the idea of the elemental originality of the sensuous, as suggested above. The expression for this idea is Don Juan, and the expression for Don Juan, in turn, is simply and solely music."[4] Later, more casual references to diabolism are comparably enthusiastic, such as Max Kalbeck's encomium to the "thrilling demonic charm of [Don Giovanni's] seductive appearance" or Alfred Heuß's praise of "the demonic element" in Mozart's music generally.[5] Like Kierkegaard, both regard demonism as essentially a good thing, a kind of electrical current that powers the highest achievements of character and composer alike.

While also inclined to see Don Giovanni as a vessel of grand forces, a second group of writers took those forces to be more worldly in origin. In a lengthy interpretation published contemporaneously with *Either/Or*, Alexandre Oulibicheff kept the character firmly among the living: "The constitution of Giovanni . . . harmonizes admirably with the material world, which for him becomes immediately the only true world, because his desires never once exceed its limits and his faculties respond to its desires. The world, nothing but the world, but the whole world!"[6] "Fin ch'han dal vino" summons an intense but mundane frenzy, a "dizzying bacchanal, an orgy of amorous exploits bereft of reason and breath."[7] Along similar lines, Carl Hermann Bitter hears a "wild high-spiritedness" in the number, and Heinrich Bulthaupt, "a wild, fresh hymn of life."[8] Victor Maurel's description of 1897, with its "procreator" Don Giovanni bent on propagating the species (see chap. 2), put the same idea in evolutionary terms, and not for the last time. In 1924 the Mozart biographer Hermann Abert saw the character as driven by an "elemental sensual urge to live and love," a formulation reminiscent of the philosopher Henri Bergson, who hypothesized that organisms were driven to adapt by a mysterious "élan vital."[9] Abert does mention demons, but they are metaphors in an interpretation that further distances the libertine from religion and metaphysics: "The concept of radical evil is unknown to the wholly unphilosophical Mozart, and so he does not judge his hero and the latter's actions by some moral ideal but feels very clearly that a power as real as Don Giovanni's life force can be defeated only by a more powerful reality: one demon can be conquered only by another. In short, the work is not about guilt and retribution but simply about being and non-being."[10] Once again "Fin ch'han dal vino" is said to be uniquely revealing, in this instance of the hero's élan: "The music not merely describes or clarifies the excitement with greater or lesser immediacy, it actually *is* that excitement. We are directly assailed by the whole pressure of the life force in full spate, a force which, freed from all constraints, races past us like an unbound force of nature."[11] Drinking, dancing, and sex, the actual subjects of the aria, are supplanted by the brute realities of the biological world.

As Abert well knew, with this reading he departed not only from Hoffmann and Kierkegaard, but also from Otto Jahn, whose own biography of Mozart (1856–59) served as Abert's starting point. Jahn took yet a third view, arguing that *Don Giovanni* is after all an opera buffa, whose protagonist has neither "a demonic enthusiasm for seduction" nor "a more powerful, soul-gripping passion."[12] He is just an exuberant lover, and "Fin ch'han dal vino" is scenic rather than sinister: "Occupied with the representation of Don Giovanni's ball, Mozart has him sing a quite simple dancing song of very lively character, in which nothing can be detected of higher passion, still less of demonic desires, nothing of nobler feelings. Instead it expresses, with incendiary power and in a very insistent and successful way, the sensuous excitement of the moment, like a quickly passing euphoria."[13] This more down-to-earth Don Giovanni attracted his own circle of devotees. Like Jahn, Ferruccio Busoni couched his opinion as a corrective: "The wrongly named 'Champagne Aria' . . . has nothing 'bacchanalian' about it. Urbane nimbleness, carefreeness, and a sparkly love of life ring out of its springing rhythms, its bright tonality. Even a marked *un*sensuousness stands out conspicuously."[14] Charles Gounod said much the same thing, calling Don Giovanni a "votary of pleasure" and concentrating on the accents and rhythms that give the aria its "striking character of energy and ardour."[15] That the ardor is for conquest goes unmentioned, as is also true, not surprisingly, in contemporary record advertising. "The gay *Don* sings his brilliant *Drinking Song*, famous in every land," says *The Victor Book of the Opera*, which offers neither a recording for sale nor a translation (see chap. 1).[16] English-speaking consumers were evidently to be protected from the aria's paean to sexual adventurism.

Another text with formidable influence among English-speaking readers, Edward Dent's *Mozart's Operas* of 1913, sums up this line of thinking.

Dent reiterates the buffa nature of the opera and offers a notably dry analysis of the hero: "The character of Don Juan falls into two easily separable aspects, Don Juan the profligate and Don Juan the blasphemer."[17] He scarcely mentions "Fin ch'han dal vino" (perhaps its "profligacy" was too obvious for comment), and when he reaches the final meeting of Don Giovanni and Donna Elvira, the only other place where the character voices anything like a personal creed, he says only: "There is a delightful exuberance about his shouts of 'Vivan le femmine!'"[18] Neither demon nor biological force, Dent's libertine is simply a merry degenerate whose antics make good comedy.

Beyond Cocky Rascals

All the agonizing over Don Giovanni's true nature has echoes in early "Fin ch'han dal vino" recordings. Mozart's music bears the fastest tempo indication in the opera, Presto, and with the vocal phrases spilling out one after the next and the rhythms suggesting quick-time marching as much as bacchic dancing, it is easy to grasp why so many accounts celebrate the number as pure energy. Yet just as writers differed as to whether the energy was demonic or comic, singers disagreed as to whether the music was relentless or playful. At one end of the spectrum lies Francisco D'Andrade (b. 1856), a prominent Don Giovanni of the turn of the century who holds a special place in the history of the aria. Among his admirers was the artist Max Slevogt, whose portraits of the baritone in character became part of the opera's lore, particularly *Das Champagnerlied* of 1902 (*The Champagne Aria*, fig. 3.2).[19] A favorite for advertisements and album art to this day, it pictures an aristocratic masculinity no less confident than Siepi's in the Salzburg production pictured above. The clothing is more decorative, but the jutting facial hair, prominent sword, and studied nonchalance all signal that under the finery lies a potent élan, as Bergson might have said.

In performance that masculinity could apparently become aggressive, as on the occasion of Slevogt's first encounter.

On a theater night in Munich in 1894, a Portuguese singer was supposed to appear for the first time as Don Giovanni. The conductor had not himself led the rehearsals, and in the Champagne Aria, accustomed to the unhurried tempo of [the German baritone] Eugen Guras, he cued the orchestra accordingly. But what was this? From measure to measure the guest set an ever wilder tempo, so that at the end the Capellmeister, who had never before encountered such masterful singing, threw away his

FIGURE 3.2 Max Slevogt, *Das Champagnerlied* (1902). Oil on canvas, 215 x 160 cm. Staatsgalerie Stuttgart. Photograph © Staatsgalerie Stuttgart. Used by permission.

baton and cried rapturously, "Da capo! Da capo!" Now it was the conductor who took a racing tempo. But Francisco D'Andrade outdid him! And there emerged a real competition between the two "cocky rascals" [*übermütigen Lausbuben*] as to who could go faster.[20]

Precise, energetic, and blazing fast, D'Andrade's recording of the aria (1906–7) largely accords with the tale of his debut. In a prefatory snippet of

recitative, he unveils a bright, nasal tone that cuts easily through supporting chords on the piano. It stands up equally well against the shrill ensemble accompanying the aria, adding bite to a performance that hovers around
145 bpm. D'Andrade maintains the tempo by applying frequent accents, by
executing the marchlike rhythms as written, and by taking the lead-ups to
the aria's three reprises without slowing down, even though they are the
only long notes in the whole number, and an obvious opportunity to stretch
the beat (e.g., "a-mo-reg-*giar* . . . Ah la mia lista"). Even the final cadence
brings no ritard, notwithstanding D'Andrade's bravura leap to an unwritten high F. Then he does it all again. After a sloppy orchestral finish (like
the conductor in Munich, the players cannot keep up), the performers or
others in the studio cry out for an encore, a make-believe "live" reaction
that triggers a second, equally blistering rendition. ♫

Rhythm and Affect

Mozart sets most of "Fin ch'han dal vino" to a distinctive five-note rhythm
(fig. 3.3a):

quarter	*eighth–*	*eighth*	quarter	quarter
long	*short–*	*short*	long	long
Fin	*ch'han*	*dal*	vi–	no

The pattern requires short and snappy articulation, especially at faster
speeds, and the composer adds further challenges by sometimes altering the
last two longs to shorts, or syncopating the first note so that it enters a beat
early. Francisco D'Andrade uses the rhythms to build momentum; though
he turns a few shorts into longs and one of the long–*short–short* sequences
into a triplet ("teco ancor"), otherwise he lands every attack as written and
with bite. By contrast, the more relaxed Mattia Battistini reduces the snap,
beginning in the first phrase where he sings most of the long–*short–short*
patterns as triplets, and the slurred eighths on "*vi*-no" and "*te*-sta" as quarter notes (fig. 3.3b). Meanwhile, Leopold Demuth falls so consistently into
triplets that his voice seems to superimpose a compound duple meter onto
the simple duple of the ensemble (i.e., 6/4 over every two measures of 2/4),
leaving a playful impression reinforced by bouncing accents on the two-
measure groups (fig. 3.3c).

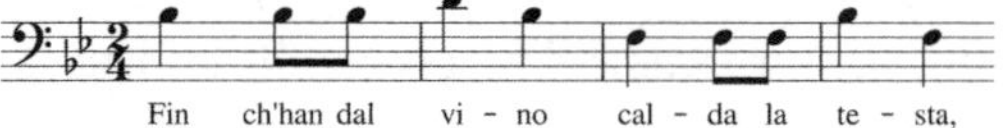

FIGURE 3.3 Rhythmic variations in "Fin ch'han dal vino"
a. Melody as written ("While they still have heads hot from the wine, have a grand celebration prepared"; Stivender and Webb, *"Don Giovanni,"* 326)
b. Mattia Battistini (1902)
c. Leopold Demuth (1909) ("Off to the party, it will be merry, for my guests love wine")

Thirty years later Ezio Pinza would recapture D'Andrade's intensity, which itself seems to channel the "wild" energy E. T. A. Hoffmann and others heard in the aria, or perhaps Hermann Abert's "force which . . . races past us like an unbound force of nature." In its own day, however, D'Andrade's reading was an outlier. Born in 1856, he belonged to the same generation that made the first round of Serenade recordings, and those of his contemporaries who attempted the Champagne Aria exercised far more rhetorical license. The 1902 version by Mattia Battistini goes nearly as fast as D'Andrade's and incorporates a similarly stagey encore. Yet the beat is less steady, the rhythmic delivery less exact, the vocal tone more diffuse (if not quite so ethereal as Battistini's "Deh vieni alla finestra"). Rather than force and wildness, Battistini projects flexibility and humor, as if the character were gripped by Edward Dent's "delightful exuberance" or Otto Jahn's "sensuous excitement of the moment." Leopold Demuth goes further in the same direction, singing approximately 15 bpm slower than D'Andrade and turning Mozart's rhythms into a kind of jig, a dance instead of a march. Across several recordings, finally, Antonio Scotti preserves the marchlike rhythms but adopts progressively slower tempos and a characteristically restrained tone: the description that comes most readily to mind is Ferruccio Busoni's "urbane nimbleness." Significantly, Scotti, Demuth, and Battistini all slow down before one or more of the aria's melodic reprises, a standard rhetorical device that tempers the aria's forward rush.

Like "Deh vieni alla finestra," the number becomes more regular as the decades pass, shedding triplets, ritards, and other liberties in favor of the written rhythms. Most later singers, however, remain as distant from

D'Andrade as Battistini and Scotti were. Several soloists who sang the aria in German take quick tempos (ca. 125–135 bpm), among them Heinrich Schlusnus (b. 1888), Karl Schmitt-Walter (b. 1900), and Gerhard Hüsch (b. 1901), but otherwise they recall the good-humored Demuth, with their light voices and dancing, jiglike articulation. Titta Ruffo (b. 1877) and Mariano Stabile (b. 1888) sing with great force, but their tempos are comparatively relaxed (ca. 105–110 bpm) and their interpretations full of nuance. Ruffo takes time to darken his voice when Don Giovanni orders Leporello to find girls for the party, then to brighten it again when the character lists off the dances they will perform. Stabile, for his part, alters course with each verse, exchanging a single burst of energy for alternations between heavy and light tone, lyrical and separated articulation. They read the aria for its details, the expressive quirks that the exegetes tended to gloss over. Abert referred to the syncopations and chromatic lines that color the list of dances as "secret forms of psychological resistance that continue to add to the sense of excitement."[21] This may ring true for a performance like D'Andrade's, but elsewhere the passage seems to disconcert the character more than motivate him: Battistini is one of several soloists who sounds off-kilter at this point (0:16–0:26). One could accuse him of sloppy execution, but he may also be imagining a less monomaniacal libertine than Abert is. A Don Giovanni who cannot keep up has more to his personality than pure energy: he may be delirious, so caught up in the moment that he loses his bearings; or he may be tired, so frenzied that he runs out of steam.

Were it not for Ezio Pinza's performances, the relentless "Fin ch'han dal vino" might well have become an antique, eclipsed by less threatening, more "Mozartian" versions. Indeed, even Pinza's studio recording of 1930 stops short of a complete revival; though it exchanges variety for modernist consistency, by the same token it sounds relatively cool and reserved, its quick tempo notwithstanding (ca. 125 bpm). Things changed when the singer began working with Bruno Walter, who had conducted D'Andrade decades before and shared Slevogt's admiration: "Another performance I shall never forget was D'Andrade's fascinating Don Giovanni, one of the rare instances where an artist seemed to have been predestined for a part by nature."[22] Intentionally or not, in the later 1930s and 1940s he and Pinza paid homage to D'Andrade by performing "Fin ch'han dal vino" at around 135 bpm and loading up the vocal and orchestral parts with heavy accents. The libertine regains his focus, pursuing pleasure with a seriousness that is only reinforced by Pinza's sole improvised effect, a concluding guffaw. This is no playful release but a contemptuous sneer, the character's "mockery of the little people around him," in Hoffmann's words. The last laugh is on those who imagine they can resist his powers.

The Athletic Don

Not everyone followed Pinza's lead: in the 1950s, successors such as Cesare Siepi and Giuseppe Taddei sang relatively lighthearted versions of "Fin ch'han dal vino" (complete with dancing triplets, in Siepi's case). Nevertheless, high-octane renderings have remained a standard down to the present, so much so that it is worth pausing over their potential relationship to contemporary modes of masculinity. While opera commentators of the later nineteenth and early twentieth centuries were gushing over Don Giovanni's free spirit, social reformers and educators were plugging a different ideal. Focused on the middle-class family, they decried libertinism and exhorted men to clamp down on their sexual impulses—literally, in some cases—the better to provide financial security for their wives and children.[23] At the same time, one of their recommendations opened the way, if not to promiscuity, then to an eroticized energy not entirely divorced from that of "Fin ch'han dal vino." Family men were told to emulate warriors, whether by joining the military or by taking part in one of the era's innumerable exercise crazes. Hardening the body, the reformers reasoned, would ward off the emasculating effects of domesticity and an ever more sedentary workplace. Unlikely as it seems, this put them in the same camp as Armand Hayem, who in 1886 lamented the decline of Don Juanism due to "small-minded bourgeois morality" and its attendant physical afflictions: "This is the century of the enervated. The nerves dominate. A nervous Don Juan is soon lost. Health is necessary for Don Juan."[24] Hayem had no investment in middle-class respectability; his study is dedicated to Jules Barbey d'Aurevilly, author of *On Dandyism and Georges Brummell* (1845), a text that helped inspire the decadence movement in later nineteenth-century France. Nonetheless, he shared reformist worries about insufficiently vigorous men, who were thought to risk hysteria and other nervous disorders associated with women. Libertines and faithful husbands alike were said to require the same vital energy, what Hayem dubs the "original élan."[25]

Other contemporary images of masculinity raise the same paradox, celebrating vitality independent of morality. As Lynne Segal shows, a prime example was the writer Ernest Hemingway, whose "work—the action-packed, concrete, laconic prose—and his life—the boxing, big-game hunting, shooting, fishing, drinking, swearing, whoring champion of any and every manly pursuit—were both crafted to teach the world the meaning of manhood: tough, patriotic, North American masculinity."[26] Opera seems worlds away, but it may be no coincidence that in the very same year, 1929, that the *New Yorker* extolled Hemingway's "profound bravery,"[27] the *New York Post* deemed Pinza's Don Giovanni to be "a roystering guardsman . . . more than . . . [a] loverly hidalgo" (see chap. 2). Meant as a criticism, the quip may equally well have registered an upheaval in opera's own ideals of manliness. On the eve of the Great Depression, which would bring its own "crisis of masculinity," the singer as much as the writer projected a newly aggressive kind of virility.

In Pinza's case, appearance may have played a role alongside vocal style. In publicity photos at least, Don Giovannis of the turn of the century look mostly like Slevogt's D'Andrade, with elaborate mustaches and embroidered doublets topped off by puffy sleeves and ruffs. The costume remained standard for decades, but images of Pinza show the finery beginning to moderate, the puffed sleeves giving way to seventeenth-century epaulettes, the ruff to less constraining collars, the mustaches to a square jaw (fig. 3.4). If contemporaneous films are any guide, the changes are significant. In

FIGURE 3.4 Ezio Pinza as Don Giovanni, from *Carnegie Hall*. United Artists, 1947; DVD Bel Canto Society BCS-D0791, 2008.

FIGURE 3.5 Images of masculinity in *Adventures of Don Juan*. Warner Bros., 1948.
a. Don Juan (Errol Flynn)
b. King Phillip III (Romney Brent)

Adventures of Don Juan (1948), for example, starring Errol Flynn, and in the similarly themed *The Affairs of Cellini* (1934), starring Fredric March, the heroes sport neat beards, plain patterns, narrow sleeves, and flat collars, as opposed to their enemies' bushy faces, elaborate embroidery, billowy sleeves, and wide ruffs (fig. 3.5).[28] Clothes convey a significant advantage: shorn of constraining ornament, Don Juan and Cellini move and speak with superior speed and force, winning in fights and romances alike. If not quite ready for Hemingway, they nonetheless epitomize a direct, physical manliness.

The 1954 Salzburg production pictured at the beginning of this chapter plays on the same contrast: in his high collar and buckled shoes, Anton Dermota's Don Ottavio is no match for Cesare Siepi's more lightly clad Don Giovanni. Pinza's costumes retained more old-school embroidery, but his Don Giovanni was sufficiently streamlined for the singer to joke about it in the only readily available clip of him performing "Fin ch'han dal vino," from the 1947 film *Carnegie Hall*.[29] A thinly veiled advertisement for classical music, the movie invents scenarios allowing famous performers to deliver their signature pieces, and Pinza's performance is as swift and forceful as ever. Equally interesting, though, is a subplot about costume. Wearing his customary bright doublet, in this instance with a fold-down collar, the singer berates a stage director who wants him to add a feathered hat (fig. 3.4). Appealing to the leading lady of the scene, Pinza cries: "Would you be crazy for me in this? No, it makes me look fat! Women will not be crazy for a Don Giovanni who is fat!" His leaner look had become a signature in its own right, fodder for Hollywood laughs.

None of this is to say that the mustachioed Don Giovannis of old lacked for masculinity. Facial hair has long-standing phallic associations, and the elegant swordsman of Slevogt's paintings conveys his own kind of menace, physical and sexual. Nor did every production of the 1940s or 1950s dress its star like Errol Flynn, any more than every singer sang like Pinza; Siepi wore confining ruffs in his time, and even at Salzburg he donned richer colors in some scenes than are pictured in fig. 3.1 (though the necklines remain open). Nevertheless, sleek outfits and athletic looks may have reinforced the structural leanings of performances after Pinza. Modernism took an implicit stance on gender when it put form ahead of emotion and complete works ahead of details. The trailing items had long been coded as feminine, and a cooler, more regular, more strictly organized art was by twentieth-century standards also a more masculine one.[30] Hemingway's writing represented the hard edge of the new style in literature, and an unornamented, unsentimental musical performance may have had a related impact within the world of opera, especially in a number like "Fin ch'han dal vino." Between 1900 and 1950, the resplendent and subtle Don Giovanni was joined by a more cinematic action hero, plainer in appearance and blunter in method.

Translating the Dandy

Questions about masculinity loomed equally large over Don Ottavio. Male and aristocratic, he seems best equipped of all the characters to bring Don Giovanni to account. Yet Mozart and Da Ponte have him take nearly the entire opera to reach a decision; to act not by issuing a challenge but by calling the authorities; and to devote his arias, both the Prague original and the Vienna addition, to illustrating his love for Donna Anna. Critics in thrall to Don Giovanni's initiative found all of this deeply unsatisfactory.[31]

Hoffmann led the way again. Describing Don Ottavio's initial entrance, just after the murder of the Commendatore, the writer verges on homophobia: "A dainty, well-groomed, spic-and-span little man of twenty-one years at most. As Anna's fiancé he was probably staying in the house, since he could be summoned so quickly. Had he hurried out at the first noise, which he undoubtedly heard, he could have saved her father, but he had to tidy up first and in any case would not have wanted to venture out at night" (18). Hoffmann goes on to charge him with frigidity, and with leaving Donna Anna to fend for herself (28). A century later Dent is similarly ill-disposed, bemoaning the fiancé's "utter inadequacy" and joking that he "follows Anna about like a little dog."[32] Others, less harsh, were still anxious. Abert sounds defensive: "Don Ottavio never grows sentimental in an effeminate way. What he feels is true, sincere, manly affection, and his weakness consists only in the fact that from his initial vow onwards he is so absorbed by this emotion that his strength of purpose and resolve simply atrophy."[33] Oulibicheff has Mozart arguing (in an invented dialogue with Da Ponte) that promises and good efforts are "all that a prima donna is justified in demanding of a loyal tenor under these circumstances"—the circumstances being that Don Giovanni is too powerful for mere mortals.[34] Kalbeck takes a more legalistic view: "According to the prevailing concept of honor, a Don crying energetically to the police makes for the most pitiable figure, and the disappointed spectator would best send him on his way with a couple of rotten apples. And yet the circumstances are in truth quite different. . . . Every honor court would agree with [Don Ottavio] when he says he cannot demand satisfaction. He no longer regards as his equal the disreputable, deeply compromised adventurer, who—as the insulted party must assume—has perpetrated an unchivalrous murder."[35] Kalbeck may be correct on the etiquette, but his resort to such an elaborate proof shows how objectionable Don Ottavio's passivity had become.

His character was further debated, and sometimes updated, in translations. As we have already seen, nineteenth-century translators had no qualms about altering Da Ponte's original, particularly in the service of social and sexual decorum. In Don Ottavio's case, the challenge was a perceived excess of decorum, especially in the character's original aria, "Il mio tesoro intanto," and its preceding recitative. The scene occurs midway through act 2, just after the sextet affords new evidence of Don Giovanni's duplicity. Convinced at last of the libertine's guilt, Don Ottavio addresses himself to Donna Elvira, Masetto, and Zerlina, Donna Anna and Leporello having already fled the stage. His recitative talks of lodging a complaint, and his aria balances a verse asking the others to comfort Donna Anna—

| Il mio tesoro intanto | Meanwhile, go and |
| andate a consolar | console my treasure |

—with another promising vengeance:

| . . . che sol di stragi e morti | . . . I wish to return only as herald of |
| nunzio vogl'io tornar | slaughter and death[36] |

The concluding line encapsulates the problem as critics and translators saw it. Whether too noble or too weak, Don Ottavio does not say he will perpetrate "stragi e morti" himself. Messenger rather than avenger, he will merely report back when the deed is done.

German translators took special pains to revise his plans, prompted in part by one of their own, Friedrich Rochlitz. In 1801 he published what would become the most popular German version by far, and his innovations included making Don Ottavio even less proactive than he is in the original. In the spoken dialogue that prefaces "Il mio tesoro intanto" (replacing the recitative), the character never mentions seeking justice: instead, he and Donna Anna—who has not fled—promise to comfort Donna Elvira in the wake of her latest humiliation. The aria continues in a similar vein, its first verse describing how "tears [will be] dried by a friend," the second offering metaphorical protection from storms. A true sentimental lover, Don Ottavio assures both of the noblewomen that "your heaven will remain bright."[37] Rochlitz's successors hastened to find more stirring language. Some cleaved to the original, restoring Don Ottavio's promise to go forth and see justice served. In Hermann Levi's text of 1897, the only one to approach Rochlitz's in popularity, the recitative restores the original condemnation of Don Giovanni (addressed to the same characters as in Da Ponte), and the aria itself preserves the substance of the Italian text, if not always its metaphorical flair. The second verse ends without violence—

Dann erst, wenn dem Verbrecher	For only when the criminal
Lohn seiner That geworden,	Has been punished for his deed,
Als Richter und als Rächer	As judge and as avenger
Werd'ich ihr wieder nahn[38]	Will I approach her again

—but it does show Don Ottavio preparing to act, promoting him from "herald" to "judge and . . . avenger."

Others remade him as a fighter. Levi was actually toning down an earlier version by Franz Grandaur, whose 1874 text he used as a model (and which enjoyed another life in Alfred Einstein's widely used Eulenburg edition of the 1940s). In Grandaur's rendering, Don Ottavio's aria is especially assertive, with a first verse pledging to protect Donna Anna with "advice and deed" (*Rath und That*) and a second waxing heroic:

Auf denn zur heil'gen Sache!	Onwards then to the holy affair!
Mein sei das Werk der Rache,	The work of vengeance is mine.
Das treue Schwert des Ritters	The true sword of the knight
Straft jeder Zeit Verrath![39]	Punishes crime at any time!

First a judge, now a knight. Kalbeck takes the latter image and sharpens it, eliminating the recitative, which he cannot abide even though he defends its social mores (see above), then changing the aria from second person to first. This allows Don Ottavio to put himself on the line:

Seh' ich entlarvt den Frechen,	I will see the cheeky devil exposed
Werde Dein Leid ich rächen;	I will avenge your sorrow;
Getrost, Dein Ritter streitet!	Be assured, your knight fights!
Er ahndet jede Schmach![40]	He punishes every ignominy!

Not content with promises, in his next appearance Kalbeck's Don Ottavio announces to Donna Anna that he and Don Giovanni will fight a duel. His rehabilitation from dandy to man of honor is complete.

Singing the Hero

On recordings, the characterization of Don Ottavio is not necessarily tied to the choice of translation. Nearly all German-language recordings use the more temperate Rochlitz or Levi texts rather than Grandaur or Kalbeck, and in any case singers go their own ways: Hermann Jadlowker gives perfectly heroic renditions of Rochlitz's weepy "Thränen vom Freund getrocknen," and Peter Anders an equally lyrical one of Levi's more forceful "Folget der Heißgeliebten." What the translations illuminate, together with the critical commentaries, is an implicit question raised by the idealization of manly action, whether in general or in its early modernist, Hemingwayesque phase. What were performers to do with a character so often faulted for passivity? Ignore the writers and embrace his sentimentality, or mind them and project a more aggressive posture, a dueling Don Ottavio to match Pinza's bold Don Giovanni? "Il mio tesoro intanto" afforded more leeway than the aria added for the Vienna performances of 1788, "Dalla sua pace," whose text and music are more consistently effusive (which presumably explains why Kalbeck cut it from his translation). In the Prague aria, the opera seria–like contrast between a consoling verse and an avenging one is matched musically by an alternation between smooth lyricism and angular declamation. Don Ottavio sings first to an *amoroso* melody accompanied by muted strings and the mellow tones of clarinet and bassoon, and all remains calm apart from a flourish of coloratura punctuating the end of the verse. But then the voice cuts to choppy and disjunct phrases, and the strings, to separated chords, syncopations, and military-style rhythmic signals (long–*short*–*short*), all played with mutes off. Compounding the shock, the harmony departs the sunny B♭ major tonic for the relative minor and eventually the dominant, where the cadence is emphatic. The remainder of the aria recapitulates the contrast, the first verse returning verbatim and the second in modified form (and the strings first restoring, then removing their mutes again). Don Ottavio gets to be *agitato* as well as *amoroso*, and his more active self gets the last word.

In navigating his mood swings, tenor soloists evidence some familiar generational tendencies. Nearly every performance distinguishes the second verse by stepping up the overall tempo, typically by accelerating toward the line about slaughter and death ("che sol di stragi e morti"), then continuing at the new speed. Tone and delivery change accordingly, gaining intensity as the tempo increases. Echoing patterns we have seen before, singers born before 1890 favor big shifts, singing the first verse slowly and flexibly, then growing more strident while also speeding up by an average of about 30 percent. Those born after 1890 set out faster, speed up less, and moderate their changes of tone, leaving a steadier impression on the large scale as well as the small. Just as the details pass without being highlighted, the sections alternate without disruption. ♫

Among earlier generations of singers, the contrasts between sections can be extreme. Recording with piano in 1907, the Italian tenor Giuseppe Anselmi (b. 1876) begins near the bottom of the tempo spectrum, around 40 bpm to the half note. He sings gently, embellishing individual words and musical junctures with slides, swells, and dramatic tempo variations, the last coordinated deftly with the pianist. Waxing and waning in both tone and speed, the first verse pulses with emotional sensitivity. And then— everything changes. In the second verse, his volume strengthens, his delivery becomes accentuated, and his tempo accelerates drastically, by some 50 percent. He appears as eager to exact justice as to convey love, pushing the forcefulness of *agitato* as far as he does the lyricism of *amoroso*.

He pushes further in the da capo, adding turns, aspirations, and an improvised rise to a high B♭ at the cadence. There his rendition ends, presumably in the interest of time, diminishing somewhat the impact of the central episode. Nonetheless the contrast is striking, and other recordings

Characterizing Don Ottavio

In both verses of "Il mio tesoro intanto," the characterizations projected by tempo and tone are reinforced by details. Tenors such as Giuseppe Anselmi and John McCormack add extra lyricism to the first verse by softening the hard consonants in the text, especially the three t's in the opening line. McCormack stretches the final syllable of that line as well, so that the written quarter note, treated by most singers as a simple phrase ending, becomes a quavering half note: "in-tan-*to*." Like many soloists born before 1890, Anselmi enunciates the beginning of the third line, "e *del bel* ciglio" (and from her beautiful eye), so that it sounds like "e *della bella* ciglio." The added syllables allow the attached musical idea, a descending scale, to sound especially mellifluous (McCormack extends the two l's, to similar effect). By comparison, Richard Tauber and others with more assertive conceptions enunciate the hard consonants of the first verse clearly and perform "del bel" as written. Similar details affect the second verse. In addition to accelerating and singing with new intensity, Anselmi also lunges at the high notes and shortens the pairs of quarter notes at the end of each phrase into eighth notes. Everything serves to ratchet up the urgency. The more moderate Tauber sings the high notes without undue stress, and McCormack, who declines to make the second verse heroic, approaches them with attenuating swells. As if to undo Anselmi's brisk cutoffs, he also softens the end of the first phrase ("torti") with an unwritten, sighing appoggiatura.

suggest how the performance may have ended had it continued. Fernando De Lucia (b. 1861) also performs a truncated version, but he cuts it in the middle rather than the end, skipping from the end of the first verse to the reprise of the second. Otherwise his reading recalls Anselmi's, accelerating by some 40 percent between the first verse and the second and ending with improvised bravura, such that the performance finishes in a rush of excitement. As performances and characterizations both, these and similar readings probe the extremes of contrast, testing the limits of what the words and score can accommodate.

A recording by Richard Tauber (b. 1891) introduces a quite different concept. Singing in Italian, he brings plenty of lyricism to the first verse, moving smoothly between notes and tapering the beginnings and ends of many phrases. Nonetheless, his tone is stronger than De Lucia's or Anselmi's, and a quicker and steadier tempo reinforces the impression of firmness, as if the character were as concerned with giving orders—go comfort Donna Anna!—as with expressing love. Vis-à-vis Tauber's predecessors, the second verse is correspondingly less forceful. Tone and accentuation do grow more powerful, but the tempo increases only modestly, and top-of-the-beat surety suggests more confidence than excitement. Tauber finishes the aria in the same spirit, incorporating accents and a rallentando into the final cadence, but nothing like the dramatics of De Lucia. His Don Ottavio stays on an even keel, tempering both the emotionality of the first verse and the heroics of the second. Many of his contemporaries strike a similar balance, though with the usual surfeit of nuance. Charles Hackett (b. 1889) and Julius Patzak (b. 1898, singing Levi's words) echo Tauber's tonal and rhythmic consistency but effect bigger tempo changes, while Max Lichtegg (b. 1910) minimizes the tempo change but increases the overall speed. Each explores a slightly different mixture of affects, but all project stability and some degree of vigor. Though lyrical in their *amoroso*, the later generations avoid languor, effectively keeping pace with the evolution of contemporary Don Giovannis toward steadiness and athleticism.

With this in mind, it is tempting to conclude that modernism finished what the translators began, salvaging Don Ottavio from his passivity by way of musical consistency and liveliness.[41] But that is not the whole story. On the one hand, if we disregard consistency for a moment and listen for affect, per se, some of the earlier generations outdo their successors in "manly" assertiveness: Anselmi and De Lucia may sound like victims of sentiment in the first verse, but in the second they become real agents of violence, as if they were singing Kalbeck's words about knights punishing ignominy. On the other hand, consistency does not always go together with assertiveness, and there are performances both early and late that aim not to balance *amoroso* and *agitato* so much as to subjugate the latter to the former. The most striking example is the 1916 recording by the Irish tenor John McCormack (b. 1884). One of the original media superstars, McCormack made his name recording ballads before establishing himself in opera, and he sings "Il mio tesoro intanto" with the same warmth that he does "Mother Machree" and "When You and I Were Young, Maggie." His light, occasionally airy tone has a noticeably relaxed vibrato, and he turns every detail of the aria's first verse to lyrical ends. More surprising, he sings the second verse the same way, as if the text and musical style had undergone no changes. His acceleration is minimal for his generation, his sound open and unforced, his delivery free of sharp edges. The accompanying ensemble cooperates, playing softly and brushing the military signals rather than punctuating them.

No one else treads quite so lightly, but several other singers come close: Koloman von Pataky (b. 1896) has a heavier tone and begins the second verse more forcefully, but his high notes are comparably delicate and his ending equally undramatic; Petre Munteanu (b. 1916) has some of McCormack's airiness and a second verse that again starts strong but relents (though in his case the orchestra appears not to notice and nearly drowns out some of his subtler entrances). Even Charles Kullman (b. 1903), who is a forceful antagonist to Pinza in the Metropolitan Opera broadcast of 1942, backs off in a test-pressing disc of 1945, rendering the first verse delicately and the second with polite heroism. Don Ottavio sings sweetly even of death, baring an unapologetically sensitive soul.

By the end of the short-playing era, then, the nature of Don Ottavio's masculinity remained very much in play. In different contexts, a single tenor like Kullman could render his biggest solo in such a way as either to confirm the character's reputation for sentimentality or undo it. Peeking briefly into the future, we can see that the debate is not yet over (we will take up the further evolution of Don Giovanni in chap. 10). In the Salzburg film of 1954, Anton Dermota's character never chafes against his winged collar, gliding through the action with somber politesse and—with Wilhelm Furtwängler conducting—lyrical deliberation. In "Il mio tesoro intanto," the *agitato* passages sound more noble than angry, and though he raises a finger when singing about death, he never reaches for his sword. Three decades later at the Glyndebourne Festival, Leo Goeke's Don Ottavio behaves and sings similarly, and so does Ramón Vargas's three decades after that, both at the Met and at Covent Garden.[42] Satisfied with loving Donna Anna and with standing up for social decorum, this version of the character forgoes the urge to violence, confident he will not be branded as weak. He can even become a new kind of hero. In a 2006 video from Zurich, Piotr Beczala sings a predominantly lyrical "Il mio tesoro intanto": calm tempo, smooth tone, sotto voce reprise tempering the excitement of the second verse. The stage direction, by Sven-Eric Bechtolf, introduces a group of women who listen admiringly and then lift the character onto their shoulders, bearing him away.

On the other hand, just a few years after Dermota's lyrical performance at Salzburg, Donald Grobe portrayed a more active and forceful Don Ottavio at the Deutsche Oper Berlin. The costume is less constricting, and the translation of "Il mio tesoro intanto," from a text by Georg Schünemann that dates to the Nazi era (and had its own afterlife in an edition by C. F. Peters [1941] and a Dover reprint [1974]), revises the already aggressive second verse of Hermann Levi to give the character still more agency. Where Levi wrote, "For only when the criminal has been punished for his deed," Schünemann substitutes, "Tell her that I will bring bloody punishment to the criminal" (Sagt, daß ich dem Verbrecher blutigen Lohn werd geben).[43] Grobe's tone hardens, his delivery grows clipped, and he grasps his sword in a show of decisiveness. In the 1980s Gösta Winbergh in Salzburg and Francisco Araiza in Milan effect similar changes of mood vocally (now in Italian), and they actually draw their swords, a gesture copied in many subsequent productions.[44] Don Ottavio becomes still more assertive in a 2006 Salzburg performance, directed by Martin Kušej and again featuring Piotr Beczala. In the same year that he was carried off by the women in Zurich, the tenor here matches the exceptionally quick tempo of conductor Daniel Harding with a forthright tone and brusque phrase endings. He also wears a sport coat and open-collar shirt, as opposed to the tuxedo and bow tie of the Bechtolf production. The clothes might mean less than they would have in Errol Flynn's day, but they do allow greater freedom of movement while also hinting at impatience with social norms, especially when the character throws off the coat and rolls up his sleeves. Sure enough, when the second verse comes he picks up a sword and begins menacing Masetto, Zerlina, and Donna Elvira, to their evident dismay. Two hundred years after E. T. A. Hoffmann, Don Ottavio remains a controversial figure. Some performances welcome the noble lover, but others still long for a warrior.

CHAPTER FOUR

All Too Human

Short-playing recordings of Zerlina's numbers, both her arias and her duet with Don Giovanni, summon another long-running debate about character. Engrossed by the libertine and his noble foes, some nineteenth- and early twentieth-century commentators paid scant attention to his peasant victim: E. T. A. Hoffmann barely mentions her, and for Victor Maurel, she serves only "the development of the character of the hero. . . . The authors did not endow Zerlina with any distinct character traits."[1] Others found that she possessed plenty of distinct traits, just not the right ones. Portrayed in the libretto as susceptible, manipulative, and erotic, Zerlina offended Romantic and Victorian sensibilities, and a great deal of effort went into sanitizing Da Ponte's words or arguing that Mozart purified them with music. By the time record companies such as Edison and Victor began marketing Zerlina as an ingenue, the girl who sings "Chide me" instead of "Beat me" to the jealous Masetto (see chap. 1), they had a century of precedent to draw on.

Much as in previous chapters, the recordings themselves tell a more complicated story. Some do suggest an artless Zerlina, as simple and innocent as the advertisements would have it. But others foreground the coquetry and sensuousness that bothered her critics, and still others breathe a solemnity that neither critics nor defenders ever imagined. More than writers or marketers, performers seem to have recognized the real perils faced by the character, and by women more generally in a culture that effaced libertinism by—among other things—treating *Don Giovanni* as a picaresque comedy. Zerlina's records suggest her pushing back, or else suffering the consequences of her impossible position in the story, caught between philandering aristocrat and jealous, powerless fiancé. More than a naïf, she sounds like a fighter, a lover, or occasionally a martyr, a character who has more in common with the New Women of the early twentieth century than with the contented housewives pictured in the gramophone advertisements.

Moral Dilemmas

Discomfort with Da Ponte's Zerlina appears as early as Friedrich Rochlitz's German version of 1801, which does its best to burnish her morals. When Don Giovanni initiates his advances in act 1, the translator has her react more forcefully than in the original,[2] and he divests the ensuing duettino of many of its sexual overtones. In Da Ponte, Don Giovanni affects an inviting first-person plural—

DG	Là ci darem la mano,	There we shall give each other our hands,
	là mi dirai di si;	there you will tell me yes;

—while Zerlina wavers:

Z	Vorrei e non vorrei,	I would like to, and not like to,
	mi trema un poco il cor;	my heart trembles a little;[3]

In Rochlitz, by contrast, the seducer commands and Zerlina expresses only foreboding:

DG	Sei ohne Furcht, mein Leben;	Have no fear, my life;
	Komm in mein Schloss mit mir!	Come to my palace with me!
...		
Z	Nein, nein, ich darf's nicht wagen,	No, no, I must not risk it,
	Mein Herz warnt mich davor;	My heart warns me against it;
	Fühlt man's so ängstlich schlagen,	When it is felt beating so fearfully,
	Hat man 'was Böses vor![4]	Something evil lies ahead!

Ultimately she cannot resist, but Rochlitz's concluding stanza implies that she has been genuinely duped by Don Giovanni's promise of marriage. At this point Da Ponte lays on the ambiguity again, having Zerlina adopt the first-person plural and join Don Giovanni in a suggestive pledge:

both   Andiam andiam, mio bene,   Let us go, my beloved,
        a ristorar le pene   to soothe the sorrows
        d'un innocente amor.   of an innocent love.[5]

Rochlitz's lovers stay in the singular and exchange pseudo–wedding vows:

both   So dein zu seyn auf ewig!   So to be yours forever!
        Wie glücklich—o wie selig—   How happy—oh how blessed—
        Wie selig werd' ich seyn![6]   How blessed I will be!

In the German, Zerlina has not "fallen" but rather believed in a lie, much as Donna Elvira talks of having done when she describes her own assent to a sham marriage (in the act 1 recitative, "Chi è là?—Stelle! che vedo!"). It would prove a popular alteration throughout the nineteenth century, surviving into Hermann Levi's 1897 translation even as he restored some of Zerlina's ambivalence earlier in the duet. Still more upright is a widely used French version by Émile Deschamps and Henri Blaze de Bury that waxes religious at both beginning and end:

DG   Là devant dieu ma belle,   There before God my beloved,
       viens me donner ta foi   Come pledge me your faith
       . . .

both   Allons, sa voix m'appelle   Let's go, his/her voice calls me
       Jurer dans la chapelle   To swear in the chapel
       Un innocent amour[7]   An innocent love.

The peasant succumbing to desire becomes a maiden being led to church.

Translations of Zerlina's arias tend in the same direction, especially "Batti batti, o bel Masetto." In a perennially controversial passage, Da Ponte has the character asking for physical punishment:

Batti batti, o bel Masetto,   Beat, beat, your poor Zerlina,
la tua povera Zerlina: oh handsome Masetto:
starò qui come agnellina   I shall stand here like a little lamb,
le tue botte ad aspettar.   waiting for your blows.

Lascerò straziarmi il crine,	I shall let you pull out my hair,
lascerò cavarmi gli occhi	I shall let you tear out my eyes[8]

In recent years graphic stagings have highlighted the violence of Da Ponte's imagery, but as we saw in chapter 1, the long-lived 1817 English translation by Isaac Pocock sought to conceal it instead, replacing "beat" with "chide." Along similar lines, Rochlitz rendered "beat" as "scold" (*schmäle*) and turned the pulling of hair and gouging of eyes into "just don't moan or complain" (*nur nicht maulen nur nicht grollen*).[9] He and other translators also toned down the double entendre in Zerlina's concluding plea, "in contenti ed allegria, notte e dì vogliam passar" (in delights and gladness let us pass the night and day).[10] Rochlitz and Levi kept the gaiety but relocated it to the daytime (*Tag für Tag*) or, more vaguely, to "the course of our young lives" (*unser jungen Lebens Lauf*),[11] while Pocock kept "night and day" but introduced a new pledge: "Here I vow to err no more." No one wanted to have an unmarried woman singing about happy couples at night.

Unable to change the text itself, critical commentators instead credited Mozart with elevating its message. Alexandre Oulibicheff interprets all of Zerlina's solos in this light, most elaborately her act 2 aria, "Vedrai carino," his favorite of the three but also—lamentably, in his view—the raciest. To console Masetto after he has been beaten by Don Giovanni, Zerlina invites him to touch her own beating heart. Imagining a dialogue between librettist and composer, Oulibicheff has Da Ponte reveling in the sexual undertones: "The little woman knows a remedy that will heal him instantly. . . . You can guess the cure, heh, heh, heh—the situation is a little risqué."[12] But Mozart responds that even "smut" (*gravelure*) may serve a higher cause: "The text is the language of the peasant; it suits him; [but] the music will be his soul, the soul of Mozart when he led Constanza to the nuptial bed."[13] The latter thought develops into a full-blown Romantic manifesto, a claim that music banishes the whole business of remedies and beating hearts. Now writing in his own voice, Oulibicheff declares: "There is no more Zerlina, no more Masetto. Something infinite, absolute, truly divine, manifests itself in the soul. . . . The aria becomes a true hymn of love, source of life and eternal rejuvenation for the whole of nature; [a hymn] to love, the springtime of souls and the most liberal manifestation of the all-goodness of the creator. It is a nuptial song for all who love, conceived in the same spirit of universality as the Ode to Joy of Schiller."[14] The final sentence undoubtedly means to reference Beethoven as much as Schiller, likening Mozart's aria to the very touchstone of musical transcendence, the setting of the "Ode to Joy" in the Ninth Symphony.

Oulibicheff's successors focus on characterization. For the biographer Otto Jahn, Zerlina's encounter with Don Giovanni features "no pathos, no hysteria [*Überspanntheit*] . . . but rather a warmly felt happiness swaying in the bright sunshine."[15] Once brought to her senses, she is innocent and charming, and the composer downgrades Da Ponte's "lasciviousness" to "a certain roguishness."[16] For Hermann Abert, Mozart intervenes more decisively. Keen to rid *Don Giovanni* of moral and philosophical overtones (see chap. 3), Abert extended his efforts to the area of intimate relations: "*Opera buffa* has absolutely no time for the complex psychological problems bound up with our sex lives but sticks instead to naked instincts, indifferent to the moral consequences of satisfying those drives."[17] Presumably aimed at Freudian exegetes, this confident assertion finds its proof in Zerlina, who is "simply an untutored peasant girl with a lively temperament, natural charm and, above all, strong, healthy instincts. It is these that determine her actions and feelings, making it impossible for us to judge her by the higher moral standards of guilt and innocence."[18] Read against the literature of seduction, she sounds like one of the lower-class girls whom Ovid or Giacomo Casanova associated with easy amusement, giving the passage a familiar voyeuristic air. What is more, Abert seems to grant her behavior—and that of her seducers, one assumes—a free pass, thanks to her supposed liberation from conventional morality. But in the next breath he changes course, asserting that the composer blocks any such sordid imaginings: "Mozart's music [in the duettino] divests this natural, entirely unreflective sensuality of all its earthly trammels, thereby granting it artistic legitimacy. By ridding the scene of the all-too-human aspect [*Allzumenschliche*] that is normally a source of pleasure to lesser minds, Mozart approaches it from the standpoint of a higher reality and creates a slice of the most immediate and natural life that silences all reservations of a moral and, at the same time, dramaturgical kind."[19] Abert uses the same language in his discussion of "Vedrai carino," leading one to wonder if his enthusiasm for sexual instinct is, after all, a rhetorical strategy meant to highlight Mozart's success in suppressing it.[20]

Whatever the case, he was not the only one to muse about alternative standards of behavior. The dramatic theorist Heinrich Bulthaupt wonders at "how far beneath [Donna Elvira] little Zerlina is morally," offering as evidence the duet from act 2, "Per queste tue manine," in which she binds Leporello to a chair and threatens him with a razor.[21] Added for the 1788 Vienna performances, the number is more typically regarded with embarrassment, whether because two buffos milking the audience for laughs falls short of expectations for Mozart, or, as Kristi Brown-Montesano suggests, because commentators and audiences feel "discomfort with a tough-talking, blade-wielding Zerlina."[22] Against all odds, however, the scene inspires Bulthaupt to Oulibicheffian visions of transcendence: "But that is the greatness of Mozart's art, that it illuminates amidst all else and with the brightness of the sun even this personality [Zerlina's]. She is so completely clad in the colors of joy, and so enveloped by a veil of grace, that we can indulge only one sentiment: that beauty is to be found neither moral nor immoral, but beautiful and only beautiful."[23] His evocation of Mediterranean sunniness recalls the light flooding through Slevogt's *Das Champagnerlied*, only here it washes away sin, much as in Jahn's description quoted above. Zerlina glows so brightly that our attention turns from her actions to the glow itself.

Inverting this image, the philosopher Hermann Cohen portrays her as the victim of Don Giovanni's glowing beauty, which raises a further moral conundrum: "Grace in its innocence must be subdued by the superior strength of personality and the magic of masculine beauty. There is no aesthetic means against it; moral doctrine certainly has means, or should always have them to hand, but if the mystery of personality in people were tailored only to ethics, there would at the very least be no art."[24] Were either character to conform to respectable standards, it would limit the opera's realism, its embrace of difficult truths about desire: "This is the highest tragedy in human life, that beauty can become a snare. There is no other aesthetic means against the power of passion than the fear of it."[25] Cohen believes Zerlina to be unaware of this, though presumably the more sophisticated characters, not to mention audiences of the opera, must face the paradox that to live fully, even good people need to fall. His fantasy of Zerlina is not one of sexual availability, but of an innocence so pure that no lapse can stain it.

Cohen's image of a vulnerable Zerlina distinguishes him from other commentators, and it finds few echoes onstage before the later twentieth century, when the emotional costs of her experiences in the plot begin to draw more attention. His determination to preserve her innocence, on the other hand, is entirely in keeping with tradition. He has her seduced by "beauty" rather than erotic attraction and kept ignorant of her transgression and its ramifications. The opera's nineteenth- and early twentieth-century commentators, nearly all of them male, simply would not accept an ambivalent Zerlina, much less a naughty or immoral one, and so they remade her as a virtuous maiden, or a transcendent angel, or a vessel of "natural" forces that render her beyond moral judgment. In this sense she recalls the demonic and biological Don Giovannis we met in chapter 3, who are likewise more symbols than characters, except that Zerlina also lacks any modicum of initiative. According to her interpreters, her good character depends on instinct, sunshine, Mozart's music—anything other than conscious action. To remain innocent, symbolic femininity must be passive.

Singing Coquetry

In speaking so confidently about the character of Zerlina, commentators may have been emboldened by the musical consistency of her two arias and duet. As Wye Jamison Allanbrook observes, "All three inflect for some time a pastoral affect," which can be heard in singable diatonic melodies, prominent woodwind parts, (mostly) simple melody-and-accompaniment textures, and swaying 2/4 and 6/8 meters.[26] The three numbers also have similar forms, with lyrical first verses, contrasting second verses, and more or less varied reprises (i.e., ABA′, or ABA′/B′)—all followed by codas featuring dancelike rhythms. Even the tempo markings are related: andante, andante grazioso, grazioso. In the eighteenth century such music would have identified Zerlina with the peasants and servants who occupy the lower rungs of opera buffa's imagined societies. In the early twentieth it would further have evoked the world of romantic balladeering that so eagerly adopted Don Giovanni's "Deh vieni alla finestra," itself a tuneful number in 6/8 with pastoral overtones. On a disc of "Batti batti, o bel Masetto" sung by Anna Maria Guglielmetti, the second side features not an aria by Mozart but the popular "Plaisir d'amour," another eighteenth-century love song in 6/8 time.[27] To an extent, Zerlina on record became a sentimentalist.

But only to an extent. Recordings of her music are also dramatic, more akin to Don Ottavio's "Il mio tesoro intanto" than to "Deh vieni alla finestra." The duettino is an active dialogue, of course, but the arias are eventful in their own ways, especially "Batti batti, o bel Masetto," which has an implied story line and music to match. It opens with some of the Serenade's tunefulness and conscious simplicity, marking Zerlina's social status and her intent to persuade. Regular, songlike phrases unfold over tonic–dominant harmonies, and a flowing accompaniment emphasizes the protestation of humility, "I will stand here like a lamb," over the beating. Yet the opposite is true of the second verse, especially at the beginning, where a back-and-forth between freshly animated strings, winds, and voice lends force to the renewed invitation to violence: "I will let you pull out my hair." An emphatic cadence follows as the character looks forward to kissing Masetto's hands, then a varied return of the first verse. At the end, finally, after a change in time signature from 2/4 to 6/8, the disputed lines about "night and day" gaiety alternate tunefulness with bursts of coloratura. From lyricism to animation to dancing and display, the aria outlines a miniature plot of its own. ♩

Its dramatic potential is evidenced by a famous disc from 1905 featuring the soprano Adelina Patti. Born in 1843, Patti was of Victor Maurel's generation and practices a comparably rhetorical style, cutting a clear trajectory through Mozart's contrasts and adding a few of her own. She begins the aria (transposed from F down to E♭)[28] with quiet pathos, signaled by a relatively slow tempo, retiring tone, and sighing portamenti. Adapted for piano, the accompaniment is equally restrained—not only the flowing figures originally assigned to the strings and winds, but also the running sixteenths heard in a solo cello, which continue throughout. An active counterpoint in some recordings, here they are a murmuring background, reinforcing Patti's almost mournful affect. The second verse brings a shift, the voice and piano both growing brighter and more rhythmically accentuated (especially through a passage of leaping sixteenths and the ensuing cadence, all preceded by a pointed appoggiatura on "gli *oc*-chi"; 0:58–1:28), and when the first verse returns it is much enlivened, including by an unwritten stringendo halfway through (1:50–2:07). At the coda, finally, Patti and her accompanist nearly double the pulse, from quarter note = circa 50 bpm in the 2/4 to dotted quarter = circa 88 bpm in the 6/8.[29] Such drastic acceleration is not uncommon in early recordings, and it may have been inspired by an editorial tradition of printing "Allegro" at the top of the 6/8, a marking that originates in sources dating to before 1800[30] (though it is not in Mozart's autograph score, a point to which I shall return). In Patti's vision, a confident, almost frothy excitement dispels the last vestiges of pathos.

Like other women celebrities then and now, Patti contended throughout her career with some of the same moral concerns that shadowed the fictional Zerlina. She managed her image accordingly, both onstage and off. As Roger Freitas writes, "Patti adopted a time-honored and 'honorable' approach to feminine acting in which all characters exhibit a 'carefully cultivated quality of innocence' that converts them more or less into ingénues."[31] Her Zerlina was a good example, garnering repeated praise when she first assumed the role in the 1860s. In the "Batti batti, o bel Masetto" recording of four decades later, Freitas hears echoes of an avowedly respectable characterization, notably in the first verse, with its gentle start, muted dynamics, and numerous examples of what I have been calling rhetorical effects.[32] At the same time, he quotes several nineteenth-century observers who recognized another side of Patti's interpretation,[33] what the voice teacher, critic, and later *Gramophone* contributor Herman Klein would remember as "a mixture of coquetry and flattery obviously meant to coax Masetto into forgiving her for her flirtation with the amorous Don."[34] Innocence aside, Patti's Zerlina has her own arts of manipulation, less sinister than those of the libertine but no less intentional. Freitas interprets the animation and accentuation in the second verse as suggestive of mischievousness and dissembling;[35] along similar lines, I hear the rollicking rhythms of the coda as a calculated effort to distract Masetto from his doubts. Put another way: Patti's reading may not be so innocent after all, at least not by the standards of Zerlina's would-be defenders. Though Oulibicheff recognized a "luxury of coquetry and of feminine solicitation" in this aria, he associated it strictly with the orchestra and especially the "sinuous" cello. The vocal melody shows only "little innocent Zerlina" petitioning Masetto "in the most ingenuous tone," a description that is hard to square with Patti's rhetorical tour de force.[36] Abert couched a similar thought in chauvinist stereotypes, writing that Zerlina "is neither a coquette already gone to seed (as so many performers seem to think) nor one of those village beauties that delighted our novel-reading, romantic grandmothers."[37] Turning the stereotypes on their heads, one could in fact imagine Patti's version of the character as both a canny manipulator and a sentimental heroine, ironic and sincere all at once.

None of the soprano's immediate successors invested "Batti batti, o bel Masetto" with quite so much drama; in the opening section especially, singers such as Marcella Sembrich (b. 1858), Luisa Tetrazzini (b. 1871), and Selma Kurz (b. 1874) affected more consistently retiring or sorrowful tones. On the other hand, certain details do suggest a more assertive teasing, such as Tetrazzini's crisp separation of the two notes on "Zer-*li*-na" at the very beginning, or Kurz's notably light articulation of the second verse (reinforced by a change from slurs to staccato in the woodwinds). Most striking, all three singers treat the 6/8 coda as an extroverted climax, taking tempos of 80 bpm or faster and in some cases improvising flourishes at the end. If Patti projected a mixture of passivity and agency, these soloists more fully embrace their own virtuosity. After a brilliant execution of the coda and a heroic prolonging of the final cadence, both Tetrazzini and Tetrazzini's Zerlina seem altogether aware of their powers of persuasion.

Claiming the Sensuous

The "Batti batti, o bel Masetto" of another celebrated Zerlina, Geraldine Farrar (b. 1882), suggests a different kind of confidence. Two generations younger than Patti, she may have felt freer to portray a more knowing and active femininity: describing a 1908 performance in Boston, the critic Henry T. Parker writes admiringly of her "archness" and "of the light, swift, bubbling, 'touch-and-go,' lyrical comedy that she rightly conceives Zerlina's scenes in 'Don Giovanni' to be."[38] Coquetry raised no qualms, and to judge from Farrar's recording of the same year, neither did a newly pronounced sensuousness. Although her initial tempo falls into the same relaxed range as that of her generational predecessors (ca. 47 bpm), her opening phrases unfold with greater evenness and warmth. She avoids volume changes, portamenti, and other strategies for differentiating individual words, and where Patti sings softly and with little vibrato—further signs of innocence, in Freitas's reading—Farrar projects fully and vibrates richly, especially at phrase endings.[39] The accompanying instruments heighten the mood, notably a clarinet whose mellow timbre substitutes for the edgier tone of the solo cello. In the second verse, the singer momentarily changes course, adding clipped appoggiaturas to "*cri*-ne" and "*gli oc*-chi," but the following sixteenths waft out in a single undulating line, and a series of generous ritards, an inheritance from earlier generations, serves mainly to stretch out the prevailing lyricism (e.g., 1:13–1:31). The same can be said for the 6/8 coda, which accelerates only to about 65 bpm and remains flowing and smooth. It offers more song than dance, and still more when the final measures slow down so much that the effective pulse shifts from the dotted quarter to the eighth. The aria finishes as it began, in a pastoral repose bordering on languor. ♫

The musical interpretation accords with Farrar's opinions on costumes, which she, like other singers of the era, largely created for herself. "I have ideas about Zerlina," she told the *New York Times* in 1908; "I shall not wear the usual velvet and spangles. My costume is of olive green cloth, very pretty, but not too expensive for that sort of a girl to wear. My hair hangs in two braids, and my skin is a dark olive."[40] Her comments prefigure the before-and-after photos published in *The Theatre Magazine* in 1911, in which she exchanges "the Zerlina costume of tradition," a plain skirt and blouse, for an embroidered dress with shawl and fan (fig. 1.2).[41] In both instances, the new Zerlina exhibits a more conventional pastoralism along with a strong dose of ethnic stereotyping, the "dark olive" skin of the *Times* description being matched by the "Spanish" ornateness of the new dress. In the photos she also strikes a more suggestive pose. Erect and scarcely adorned, eyes fixed on some unseen event, the "before" Zerlina looks like a classic comic servant, unpretentious and poised for action. Her alter ego looks more like a lover, exotic and tempting with her reclining posture, tilted head, and gaze trained on the camera. Still subject to the strictures of respectability, Farrar nevertheless carved out new freedoms, abandoning some aspects of Victorianism for the less formal fashions and comportments of modernity.

Well aware that her innovations posed risks, Farrar argued in 1909 that any offense was unintentional: "I am essentially sensuous, but have a horror of vulgarity. *Suggest* all you will, but don't *be it*; but sensuous I am in love of color, line, sound, and thought, as well as appeal. Zeal and high spirits have often evoked rebuke along the lines of the very thing I have tried to avoid. I believe I *conveyed* all right, but its receiver was not primed for that *quality* of ammunition."[42] Given her popularity, there were presumably many who felt inspired, rather than threatened, by a woman bold enough to call herself "essentially sensuous." Yet even they could react with ambivalence, putting new constrictions on the singer and her characterizations. Though he published Farrar's reflections—in a book entitled *Success in Music and How It Is Won*, no less—the writer Henry T. Finck was not only a commentator on music but a critic of the turn-of-the-century women's movement, which he feared would masculinize women.[43] This puts his admiration for Farrar in a new light, including his defense of her New York performances of *Don Giovanni*: "Some thought she over-acted, but is it not quite natural for a girl in Zerlina's situation to over-act in the exuberant outpouring of her feelings?"[44] This is no defense, really, but an embrace of sensuality as something proper to what Finck called "womanly women," an appealing subject for the male gaze.[45]

Whether or not they shared Farrar's convictions, several of her contemporaries sang "Batti batti, o bel Masetto" with comparable warmth and relaxation. Though her frequent portamenti recall Patti, Lola Artôt de Padilla (b. 1876) favors a liquid tone full of vibrato, and she sustains a lyrical mood throughout the aria, aided by Rochlitz's less piquant text. Her coda develops more energy than Farrar's, but it ends quietly and without the excitement displayed by singers of an earlier generation. Ninon Vallin (b. 1886) and Lucrezia Bori (b. 1887) are similarly calm and consistent, and so is Elisabeth Schumann (b. 1888), who recorded the aria several times in both German and Italian. Always more lyrical than comic, she became more so with the passing years, and her version of 1926 (in Italian) features one of the most legato performances outside of Farrar's own, and one of the slowest codas. At the dawn of the electric era of recording, "Batti batti, o bel Masetto" was as much romantic idyll as witty flirtation.

Gentle Wit

In the remainder of the short-playing era, both "Batti batti, o bel Masetto" and "Vedrai carino" reach a kind of middle ground between Patti's coquetry and Farrar's romance. A 1934 radio broadcast of "Batti batti" reveals a continuing urge to protect feminine virtue, as a spoken introduction by the soprano Rosa Ponselle (b. 1897) erroneously describes the character as married: "Zerlina is heard coaxing her husband Masetto to forgive her innocent flirtation with Don Giovanni. The kisses at the end prove that her feminine charm has triumphed."[46] Ponselle goes on to a brilliantly rhetorical performance, full of tempo variations and word painting used mostly to droll effect.[47] It is a late tribute to Zerlina as a clever extrovert. More typically, the shift to electric recording and toward modernist steadiness led to a remixing of wit and lyricism. So often muddy in earlier years, Mozart's intricate texture emerges more clearly on electrically recorded discs, foregrounding the interplay between voice, ensemble, and noodling cello. If the singing is not so continuous as Farrar's, moreover, soloists such as Adele Kern (b. 1901), Bidu Sayão (b. 1902), and Irene Eisinger (b. 1903) tend toward smooth delivery and comparatively subtle evocations of wit. All three, for example, single out the word "agnellina" in the opening line ("ohne Klage," in the Levi translation), but unlike Patti and others, who did so with portamenti or rallentandi, they use simple volume, treating it as an echo of "starò qui." Similarly, in the second verse, rather than using appoggiaturas or tempo variation to articulate the images of the hair and eyes and kisses, they call on volume, light accents, and even lighter rubato. As for the coda, their average acceleration from 2/4 to 6/8 is only about 40 percent, as opposed to the 60–120 percent of Patti, Kurz, or Tetrazzini, and the resulting pulse falls into a moderate range of around 73–77 bpm. Neither a stretto nor a slow pastoral, the closing takes on a lilting quality, its dancing more graceful than excited. ♫

In Zerlina's other solo, the intoned repeated notes in the main theme ("Ve-drai ca-ri-no" [You will see, dear one]) and the velvety sounds of horns and clarinets long inspired a deliberation bordering on solemnity. Artôt de Padilla's rendition of 1915 features a very relaxed tempo of 85 bpm to the eighth note and a delivery that rolls right over the number's own opportunities for witty contrast, which include "no"s reached by leaping fifth on an awkward second beat ("e lo speziale non lo sa far, *no*" [and the apothecary doesn't know how to make it, no]) and numerous iterations of "Sentilo battere" (feel it beat) to cutesy sixteenths.[48] The word painting works equally well in the German translations, yet Artôt de Padilla and her contemporaries largely ignore it, keeping the upward leaps smooth and the "beating" sixteenths connected. They enfold Masetto in nocturnal lyricism, a tendency that persists among artists of later generations such as Sayão and Yvonne Brothier (b. 1889).

By the end of the era, however, the tradition has changed. Singers including Nadine Conner (b. 1907), Maria Cebotari (b. 1910), Elisabeth Schwarzkopf (b. 1915), and Erika Köth (b. 1925) all treat the leaping "no"s and beating hearts as witty ruptures, and their tempos divest the idyllic surroundings of any solemnity. With an eighth-note pulse around 110 bpm, 20–25 percent faster than the previous standard, Cebotari's and Köth's discs interpret "Vedrai carino" as a dance in its own right, suggestive of the moderate 6/8 and dotted rhythms of the pastoral *siciliana*.[49] The aria approaches the same middleground inhabited by "Batti batti, o bel Masetto" in the 1940s and 1950s, accelerating toward gentle wit where the other had relaxed into it. By midcentury a relatively extravagant Zerlina, coquettish or sensual or solemn, had given way to a subtler agent of persuasion.

Ecstasy

The character experiences a somewhat different fate in recordings of "Là ci darem la mano," the duettino with Don Giovanni. In some ways, the earliest versions evoke the more dramatic renditions of "Batti batti, o bel Masetto"—not surprisingly, given that the two numbers follow such similar courses: songlike first verse with regular phrases and harmonies, dynamic second verse with back-and-forth alternations, varied and extended reprise of the first verse, and dancing coda in 6/8 time. Yet a signal difference changes the outcome. The duettino's reprise (the A′/B′ in ABA′/B′) is not varied so much as remade, such that the characters no longer sing complete melodies, as they do at the beginning, but rather trade entrances of two measures, then one measure—and then they overlap. This prepares a final and fateful exchange, in which Don Giovanni twice calls "andiam" (let us go) and Zerlina assents. At this point many performers reach beyond the scene's pastoral frame to suggest a kind of apotheosis. ♫

An example may be heard on a disc pairing Eva Tetrazzini (b. 1862), sister of Luisa and in this instance a very mercurial Zerlina, with a comparatively steady Luigi Baldassare. Their pianist begins with an unwritten measure of heavy oom-pah background, giving the impression that Don Giovanni is accompanying himself, much as he does in "Deh vieni alla

---

Signs of Hesitation

In recordings of "Là ci darem la mano," Zerlina's indecision is frequently articulated by rhetorical devices. One comes in her very first entrance, where, as Wye Jamison Allanbrook notes, the rhythm already suggests "anxious reflections" inasmuch as fleeting pickups to the main beats (*"Vor*-rei e non vorrei," I would like to, and not like to) break apart the "long–*short–short–* long–long" pattern of the original melody ("Là *ci da*–rem la"; see *Rhythmic Gesture*, 264–65). Sopranos such as Emma Eames (b. 1865), Ellen Brandt-Forster (b. 1866), Emilia Corsi (b. 1870), and Irene De Bohuss (b. 1878) inject further anxiety by lifting after the first "vorrei," a practice that continues in some performances today. Eva Tetrazzini does not perform the lift, but gasping appoggiaturas in the subsequent line have a similar effect (*"mi tre*-ma," *"cor"*). A further sign appears soon after, where an expansion of Zerlina's answering phrase from four to six measures, itself suggestive of hesitation, becomes more so when Tetrazzini and others add a substantial ritard on "ma può burlarmi ancor" (but he could still make light of me). But most suggestive of all is the execution of "non son più forte" (I am no longer strong), which is repeated three times in succession at the end of the second verse, then again near the end of the 2/4 (Gronda, *Il Don Giovanni*, 22; Stivender and Webb, "*Don Giovanni*," 315). Tetrazzini, Eames, and several others all accelerate through the first two iterations of the passage and slow down on the third, creating a rhythmic analogy for the insecurity expressed in the text. The shaping of such moments makes it all the more striking when the same singers do their about-face, singing "an-*diam*" in full voice and with strong, quick appoggiaturas, then launching into breathless 6/8 codas.

finestra."[50] Baldassare sets out in the same spirit, so much so that though he sings in Italian, his full-voiced, rhythmically even, unornamented traversal of the melody seems less suited to Da Ponte's ingratiating first person than to the imperatives substituted by the German translator Rochlitz (see above). Accompanied by the same stolid plunking, Tetrazzini at first adds more gravity, but her delivery is quickly unsettled by gasping appoggiaturas, a ritard, and a quavering unwritten turn. Assured to a point, her Zerlina also betrays a note of nervous anxiety.

Much of rest of the 2/4 unfolds similarly. Apart from a stretched-out anticipation of the reprise, Baldassare sings firmly and evenly all the way through the characters' overlapping exchanges and into the "andiam"s, whose falling fifths he executes as energetic downward leaps, decisive and very nearly in tempo. Meanwhile Tetrazzini continues to mix strength and nerves right up until her assent, also on "andiam," signals a surge of confidence. It is as forceful as Baldassare's, aided by the soprano's revision of the notated stepwise descent into an echo of the baritone's falling fifth (F♯–B). The following 6/8 coda is accordingly triumphant, thanks in no small part to an acceleration that exceeds those heard in early versions of "Batti batti, o bel Masetto"—and which may be indebted, again, to an editorial "allegro" that appears in many scores. On average, couples in which at least one member was born before 1885 increase the tempo by nearly 150 percent, such that the dotted-quarter pulse of the 6/8 is up to two and a half times faster than the quarter-note pulse of the preceding 2/4.[51] Dancing gives way to a kind of whirling rapture, and Tetrazzini and Baldassare up the ante by contriving to change tempo not once but three times. Thanks to a pair of unwritten rallentandi, they slow down and speed up again and again, building a momentum that spills over into bravura prolongations of the final cadence.

Contemporaries echo key features of their performance, notably the breathlessness and bravura of the ending and Tetrazzini's audible progression from indecision to confidence. Where they differ is in the seeming agency of the characters. When the baritone sings "andiam" as Baldassare does, quickly and firmly, the coda's energy may seem to originate with Don Giovanni, or with Don Giovanni and Zerlina equally. Just as often, however, the men evoke more romantic strategies of seduction, using rhetorical devices of their own (tempo variation, portamento, weepy aspirations) to emphasize invitation over command, and singing the final "andiam"s as quavering descents or lingering fades. Examples include Mario Sammarco (b. 1868), Leopold Demuth (b. 1861), and especially Mattia Battistini

(b. 1856), who improvises a falsetto climax during an introductory scrap of recitative, then fills the duet itself with pushes and pulls, scoops and slides, commands and caresses. As mercurial as Tetrazzini, he makes Zerlina's transformation seem all the more complete, to the point that when his fellow soloist Emilia Corsi (b. 1870) declares "andiam," it sounds as if Zerlina has seized the initative. Her mind made up, the object of seduction propels a sentimentalist Don Giovanni toward ecstasy.

Ewig und Selig

Except when she doesn't. Singing with a very sentimental Ramon Blanchart (b. 1865), Lydia Lipkowska (b. 1882) ends the 2/4 with a sigh rather than an accent, pausing briefly and swelling into "andiam" before fading further down the scale than Mozart calls for (from E past the written D to B). Given the roaring 6/8 that follows and some concluding vocal fireworks, the moment seems in retrospect only a last-minute hesitation. In other recordings, however, similarly effusive "andiam"s prepare a different kind of ending altogether. Farrar's disc of 1908 features her stage counterpart at the Met, Antonio Scotti (b. 1866), and both echo the characterizations they develop in their other recordings. Scotti favors a light touch and unwritten, "aristocratic" dotted rhythms, just as in his Serenades (see chap. 2), while Farrar projects more sensuousness than either anxiety or confidence, just as in "Batti batti, o bel Masetto." In keeping with this, Scotti's lyrical "andiam"s trigger a prolonged reply—not a sigh, like Lipkowska's, but a warm assent—and the following 6/8 is notably relaxed. The tempo increases by a comparatively modest 78 percent (ca. 45 vs. ca. 80 bpm), and the soloists manage the ending without pyrotechnics. ♫

Their performance may represent more than personal leanings. Farrar had sung Zerlina at a 1906 Mozart commemoration in Salzburg (an ancestor of the famous Festival), directed by her mentor, the German soprano Lilli Lehmann. According to Lehmann, the production exposed an emerging split in conceptions of the duettino, thanks to some cavalier behavior on the part of Francisco D'Andrade, he of the swift Champagne Aria (see chap. 3): "D'Andrade/Don Juan came only to the last rehearsal. For all his chivalrous politeness, since he sang 'his tempi' in all the ensembles without troubling about the other singers, disparities unfortunately resulted that none of us managed to eliminate. For example, I could not convince my dear colleague that the traditional change of tempo marking 'Allegro' in the duet with Zerlina was false. It indicates only a change of rhythm."[52] Lehmann was appealing to Mozart's autograph score, in which the 6/8 bears no new tempo indication. D'Andrade might have retorted that the editorial allegro marking originates in very early sources, just as it does in "Batti batti, o bel Masetto." Nonetheless, that Farrar and Scotti accelerate only modestly at the 6/8 may well reflect Lehmann's influence, as well as that of Gustav Mahler, who conducted the two soloists at the Met. He had witnessed the 1906 production, and Lehmann writes of having dissuaded him from taking other passages in *Don Giovanni* too swiftly.[53] Equally striking, for a celebrated 1905 production of the opera, mounted in Vienna in collaboration with the designer Alfred Roller, Mahler caused an uproar by choosing Friedrich Weidemann (b. 1871) for the title role over the company's usual Don Giovanni, Leopold Demuth (b. 1861).[54] Weidemann's recording of the duettino, made in 1907 with Elise Elizza (b. 1870), has far more in common with the Farrar and Scotti rendition than with the 1902 version by Demuth and Ellen Brandt-Forster. In 1902 Don Giovanni is sentimental, Zerlina assertive, and the coda a breathless frenzy. In 1907, by contrast, both characters are even-tempered and warm, and the coda is even less excited than Farrar's and Scotti's.

Steadier and slower points the way toward the future. The tempo disparities between 2/4 and 6/8 never diminish quite so far as in "Batti batti, o bel Masetto": among couples with at least one singer born after 1885, the average acceleration remains around 90 percent, nearly a doubling. Yet in terms of affect, the coda trends toward calm and the preceding dialogue toward ever milder intensification. Two recordings by the baritone Heinrich Schlusnus (b. 1888) illustrate a characteristic development. The first, made in 1920 with Lotte Lehmann (b. 1888), sets off like a modernist updating of Farrar and Scotti or Elizza and Weidemann, both Lehmann and Schlusnus sounding measured and reflective. Midway through the 2/4, however, they hark back to older traditions, varying the tempo and mood and eventually rendering the "andiam"s—or rather, the German "Wohlan"s—with enthusiasm. Though not quite a delirium, the coda is accordingly quick and emphatic, another glimpse of ecstasy. Schlusnus's second recording, on the other hand, made in 1936 with Erna Berger (b. 1900), begins and ends with equal restraint: when Berger reaches "Wohlan," she sings a long sigh, preparing an almost reflective 6/8. Both recordings begin around 43 bpm, but the coda with Berger reaches only about 75 bpm, whereas that with Lehmann jumps to approximately 100 bpm.

Amidst the spreading calm, many Don Giovannis continue to effect some intensification, though on the scale of the number rather than the individual phrase. In 1936, for example, Schlusnus gradually increases his volume and tonal urgency, generating a wave of feeling that comes across all the more clearly thanks to the electrical recording process. Ezio Pinza does something similar in his 1939 recording with Elisabeth Rethberg, releasing his booming bass in slow stages. His trajectory crests at the same moment as Schlusnus's, just before Zerlina expresses her final doubts, after which both singers return to a hush. In the best modernist fashion, the men's exhortations combine into a larger whole. Meanwhile, interpretations of Zerlina's role push in two directions at once. Sopranos such as Toti Dal Monte (b. 1893) and Maria Zamboni (b. 1895) affect a growing playfulness. Dal Monte, in particular, seems to toy with the straining Don Giovanni of Augusto Beuf (b. 1887), adopting a light tone and filling her lines with inquisitive lifts. Their conclusion finds a middle ground, keeping a moderate tempo of about 100 bpm but featuring a clipped, separated articulation that makes it sound faster. All but reversing the intentions of the text, her clever Zerlina appears to draw Don Giovanni into her own musical orbit.

More commonly, Zerlinas of the 1930s and beyond become ever more grave. Favoring rich timbres and a tendency toward ritardando, Emmy Bettendorf (b. 1895) and Margherita Perras (b. 1908) approach the pathos of some of the earliest "Batti batti, o bel Masetto" recordings, and together with Gerhard Hüsch (b. 1901), with whom both recorded, they turn the coda into a muted celebration in moderate tempo. Gwen Catley (b. 1906) does something similar in her 1943 recording with Dennis Noble (b. 1898). Notwithstanding a brighter 6/8 and a cheery English translation ("With thee, with thee, my treasure / This life is naught but pleasure / My heart is fondly thine"), in the 2/4 she counters the seducer's earnest pleas with palpable tones of self-questioning. Gravest of all is a 1952 recording by Elfride Trötschel (b. 1913), who gives an impression of great reluctance by clinging to a measured quiet through even the most intense exchanges of the 2/4. Fittingly, she and Horst Günter (also b. 1913) sing one of the most restrained codas of all, with a pulse of only about 58 bpm and a surfeit of lyrical sustain. Leaving behind any hint of rapture, they transport the 6/8 to a plane of bodiless emotion.

So many of the more solemn duettinos are sung in German that it is tempting to connect their affect to the words, and especially to Rochlitz's "So dein zu sein auf ewig, wie glücklich o wie selig" (So to be yours forever, how happy oh how blessed), which was preserved verbatim by Hermann Levi. To certain singers at least, the sacred overtones may have evoked something more exalted than an opera buffa seduction, something closer to the transcendent unions of high Romanticism. If they were not aiming for *Tristan und Isolde*, exactly, they may have wanted to imbue the duettino with a less physical sense of time and place than a lilting 6/8 would suggest. Or perhaps their growing seriousness represents a return to protecting Zerlina's character, in this case by portraying her as remorseful or absent, even as she consents. In 1936, in the earliest complete recording of *Don Giovanni* made under studio conditions, Audrey Mildmay maintains a sad lyricism nearly to the end of the number. The tempos are similar to the Trötschel and Günter disc, and between Mildmay's calm tone and her sighing performance of the fateful "andiam" the character appears wrapped in guilt or reverie. The mid-century Zerlina contemplates an absent subject, the abandoned beloved or some otherworldly dimension in which love escapes mere earthly commitments.

Somber duettinos set a precedent. Having dipped once in the 1930s and 1940s, the tempo of the coda settled further in the 1950s and 1960s to approximately 60–70 bpm. It recovered somewhat in the wake of the historically informed performance movement, but today it rarely reaches beyond 80 bpm and never into the supercharged realms of the earliest recordings. Seemingly unimportant in itself, the slowing is nonetheless emblematic of the debate over Zerlina's femininity, which remains as unresolved as that over Don Ottavio's masculinity (see chap. 3). As evidenced on video recordings made since 1950, the tuneful melodies and dancing rhythms of her music have frequently gone together with a wide-eyed innocence that would have pleased the Victorians. Beginning in the late 1970s, however, a mounting darkness in Don Giovanni's character makes the duettino, especially, look anything but sunny or beyond moral judgment. On the contrary, the seduction comes across as a ruthless betrayal and Mozart's music as a dirty trick, its beauty ushering in disaster rather than ecstasy or ennoblement. The 6/8 sounds ironic at best, transporting Zerlina into an elysium that is plainly false.

An increase in more "knowing" Zerlinas raises further ironies. By recognizing her own wit and sexual allure and turning them against Don Giovanni and Masetto alike, this version of the character banishes the last vestiges of passive naïveté. Yet as Geraldine Farrar already discovered, a more sexualized Zerlina can invite a disempowering voyeurism, and she may also inherit some of Don Giovanni's depravity, playing the siren to his predator. Trading infantilization for objectification, or new forms of moral opprobrium, she remains nearly as vulnerable to social stricture as she was in the nineteenth century. And also to acts of redemption. In a

2013 video featuring Katija Dragojevic and Erwin Schrott, the one play-ing an erotic coquette in modern dress, the other a violent drug addict, the duettino affords a moment of respite against a starry nighttime sky.[55] The opening is as deliberate as it gets, around 40 bpm, and the characters sing from opposite ends of the stage, each pinpointed by a spotlight. During an equally unhurried coda, played at about 60 bpm, they inch toward one another through the darkness, halos of light converging. The Victorians would have frowned at Zerlina's eventual perching on Don Giovanni's lap, but the spacious tempos and mythic setting might otherwise have pleased them. Taking a break from gritty realism, the production casts the duettino as a salute to the transfiguring powers of Mozart. Still beautiful after all these years, his music once more delivers Zerlina from the all too human.

PART II
Invented Works

Complete Audio Recordings

CHAPTER FIVE

The Virtual Stage

Commercial audio recording changed radically in the decade after World War II, with profound consequences for opera.[1] The introduction of "microgroove" LPs, played at 33⅓ rpm and holding twenty-two and a half minutes per twelve-inch side, allowed long pieces to unfurl with far fewer interruptions than had previously been the case. Recordings of complete operas followed in a flood as the major European and American labels vied to build catalogues of the standard repertoire. Meanwhile, two further technologies revolutionized the recording process itself. One was magnetic tape, which made it possible to record on multiple tracks and to splice different performances into a single, edited ideal. The other was stereo, which expanded the implied space of recorded sound to include left and right as well as near and far. The latter developments took time to catch on, but by 1960 the industry standard for opera had shifted. After a half century and more of dominance, the excerpt record gave way to the multidisc set, a carefully engineered representation of a full-length work.

The new format would reign for about as long as its predecessor. Further innovations awaited, notably digital recording, with its unprecedented clarity and dynamic range, and the compact disc, with its uninterrupted duration of forty-five minutes and more.[2] Both gained currency in the 1980s. Yet neither posed an immediate challenge to the multidisc set, which persisted in CD form even as video discs, and eventually video and audio streaming, challenged its status as the preferred medium for recorded opera. By that time, the early 2000s, audio sets had captured several more generations of performance practice and established a whole new kind of operatic experience. If short-playing records sketched characters and vignettes, sets offered entire dramas, complete in Mozart's case with the copious dialogue performed in the speechlike style of recitative. While they could not show stage action, furthermore, LP and CD sets included picturesque sound effects along with translations, commentaries, still images, and other stimuli for the imagination. All of this now came packaged with the records, inviting consumers to lose themselves in a lengthy, richly detailed, musical-literary-visual "production." Complete opera recordings turned the spaces of listening—themselves expanded by car radios, the Walkman, and other new playback technologies—into virtual stages.

Putting Opera on a Shelf

Complete opera recordings take several forms. What we often call "studio recordings" are made under controlled conditions without audience, whether in an actual studio or in a theater or other venue taken over for the purpose. They allow maximum use of multiple takes, splicing, stereo effects, and other devices of postwar audio engineering, as well as all possible combinations of performers (at least in theory; in practice, the personnel on complete opera recordings often collaborate on a contemporaneous stage production as well). A second and related option is the live capture intended for immediate commercial release, which typically combines material from two or more theatrical or concert performances. These, too, admit of considerable engineering, notably the splicing together of different renditions—and again, those taken from concert performances may feature "dream teams" of performers who do not actually appear together in a live production.[3] Like studio recordings, they aim to present an ideal performance, free of mistakes and the contingencies of the theater.

Both types differ from what I call archival live recordings, which present single performances of a given production. Usually originating in radio broadcasts, they feature the least amount of editing and the greatest range of sound quality, and they rarely achieve the circulation of their studio or studio-edited counterparts. Most appear years after the original event, and on specialist labels geared toward collectors. The exceptions tend to involve high-profile artists who did not make studio recordings of the work in question, such as Bruno Walter and Ezio Pinza, whose 1942 Met broadcast has been reissued numerous times over the years, or Wilhelm Furtwängler, whose 1954 version from the Salzburg Festival counts among the most famous of all *Don Giovanni* recordings (see the introduction).

I have been able to consult some seventy-four complete sets of *Don Giovanni*.[4] The Discography divides them into two chronologies, one made up of studio-produced versions (forty-four), and the other of archival releases (thirty). Part II references both but emphasizes the former, as it is there that one sees the full resources of the industry brought to bear on creating virtual productions, from casting to packaging. *Don Giovanni* inspired a lot of effort early on, beginning with a 1950 studio recording by a specialist American label, the Haydn Society. Within a decade several major companies had weighed in, including Philips, Cetra, Deutsche Grammophon, EMI, Decca, RCA Victor, and Pathé (an edited live version). In addition, the one complete *Don Giovanni* from the short-playing era, made by HMV with a mobile recording truck at Glyndebourne following the conclusion of the festival in 1936, was reissued on LP. Thus by 1960, LP enthusiasts could hear a pre–World War II cast captured in its prime; an older star, Mariano Stabile (b. 1888), reprising the title role on the Haydn Society recording; and no fewer than seven additional casts and orchestras, sometimes overlapping, drawn from the highest profile international houses and festivals of the 1950s.

Following a brief hiatus, 1966 inaugurated a period of two decades in which new studio releases appeared roughly every other year. EMI, Deutsche Grammophon, and Decca all released two or three more versions, each highlighting successive generations of singers, and they were joined by single entries from Eurodisc and CBS/Sony, the latter doing double duty as the soundtrack to the 1979 feature film of *Don Giovanni* directed by Joseph Losey. Then came a second outpouring, at least thirteen new recordings made between 1988 and 1997. The occasion was the Mozart bicentennial, which coincided with the advent of the CD and the expansion of the historically informed performance movement into opera. The familiar major labels (with their corporate descendants and subsidiaries) were augmented by the digital pioneer Telarc, and the ranks of the international houses, by historically informed performance specialists such as Nikolaus Harnoncourt and John Eliot Gardiner.

In the decades since, production has returned to prebicentennial levels, and with a new emphasis on edited live versions or on productions released concurrently on video.[5] It is too early to know if this represents a decline; should complete operas prove lucrative on streaming sites, a resurgence in studio-made audio recordings is always possible. Still, for now the momentum seems to lie with video, and it may be telling that Sony chose to market its first poststreaming sets of the Mozart–Da Ponte operas (2014–16), led by Teodor Currentzis, as quasi–collector's items. Consumers who wished to have physical copies could choose between hardcover books with the CDs tucked inside, or—for the truly nostalgic—vintage cardboard boxes with twelve-inch vinyl LPs. The performance and production are cutting-edge, but the packaging harks back to the golden age of the format.

The apparent waning of complete studio recordings brings their impact into sharper focus. With LPs in general, the classical record industry upped the conceptual gravity of its product. As we saw in chapter 1, short-playing records preserved excerpts and circulated with minimal labeling, leaving buyers to organize and interpret them largely on their own. LPs and CDs are more like books: items of culture complete in themselves. They have slim spines embossed with authors and titles, conducive to library-style alphabetizing and shelf scanning. They have distinctive visual designs: album art came into its own during the 1950s and 1960s, and operas inspired their share of striking examples. And of course, they offer beginning-to-end pieces along with all the interpretive context mentioned at the beginning of the chapter. Listeners no longer need *The Victor Book of the Opera* to compensate for a missing narrative or spectacle.

Complete recordings project wholeness in other ways as well. In Mozart's case, they emerged concurrently with the new scholarly edition, and they often evince a similar interest in establishing a definitive work. Fully a third of the studio-produced *Don Giovanni*s bill themselves as presenting the Prague or Vienna versions, relegating the alternative numbers to an Appendix, just as the *Neue Mozart Ausgabe* volumes of 1968 do (see the introduction). A few others put all the music into a single expanded sequence, a solution that dates to a score published in the 1940s by the eminent Mozart scholar Alfred Einstein. Long before period instruments and historically informed performance practices became common, *Don Giovanni* recordings also "corrected" long-standing traditions with reference to the original sources—for example, by substituting B♭s for the customary B♮s in the opening theme of the overture, a revision suggested by Einstein.[6] As much as scholarly scores, complete recordings promoted the idea that a particular version of an opera could be authoritative.[7]

Paradoxically, in some quarters they gained more authority from the very thing that guaranteed their incompleteness: separation from the stage. A 1937 *Gramophone* editorial deemed the complete *Don Giovanni* from Glyndebourne to be "the nearest illusion the gramophone has yet produced of an actual opera being performed." Not content with that, the writer goes on to say that listening to the record is actually better than watching a performance:

> Most of the criticism which can be levelled against [*Don Giovanni*'s] un-
> certain dramatic construction, its occasionally obscure dialogue, and its
> irrational plot disappears in listening to a continuous playing of the forty-
> six sides. . . . Moreover, some of the scenes are positively helped by being
> beheld only with the mind's eye. We can imagine for ourselves the arrival
> of that relentless statue at Don Giovanni's supper table, picture for our-
> selves the flames of hell shooting up as he is dragged below by demons.[8]

Two decades later, the surge of LP sets inspired related thoughts. The critic Andrew Porter wrote of one example that it "is not so much a theatre per-formance of *Aida* captured on records (and admittedly it does lose some-thing through not being it), as a sound-realization of the score which tran-scends any shortcomings inherent in physical staging, and brings us a step closer to that idealized performance experienced by the expert reader of the miniature score."[9] Not quite ready to declare the recording superior to the live event, he nevertheless associates the one with ideals (and reading—see further below) and the other with shortcomings. The producer of the first *Der Ring des Nibelungen* (1958–65), John Culshaw, went a little further: "When I hear it again it is the characters of Wagner's drama who come to life, and in so doing instantly dispel any memories of artists standing in front of microphones. Hagen is Wagner's Hagen, and not just Gottlob Frick pretending to be Hagen."[10] A producer is hardly an impartial observer, of course, but Culshaw's celebration of Frick's "disappearance" is striking all the same. Rather than bringing stars into the home, as short-playing rec-ords were praised for doing, complete recordings introduced listeners to the fictional characters themselves, giving access to the work behind the performance. The philosopher and critic Theodor W. Adorno pitched the same thought still more strongly. Welcoming LPs for making "the entire musical literature . . . available in quite-authentic form," he praised their benefits for opera in particular: "The LP . . . frees itself from the capri-ciousness of fake opera festivals. It allows for the optimal presentation of music, enabling it to recapture some of the force and intensity that had been worn threadbare in the opera houses. Objectification, that is, a concentra-tion on music as the true object of opera, may be linked to a perception that is comparable to reading, to the immersion in a text."[11] Listening at home is superior because it is like reading, a metaphor that would have pleased the marketers who designed the booklike box set (if the product did not inspire the metaphor). For Adorno, complete recordings allowed opera at last to attain the status of a literature.

Picturing the Don

Needless to say, complete recordings could not really capture operas in an ideal state. If anything, the industry's push to release competing versions highlighted the variability of repertoire staples like *Don Giovanni*, and the accumulation of ever more versions since the 1950s has only reinforced the point. Beyond differences in musical or dramatic interpretation, their very packaging betrays their singularity. Both the foot-square covers of LP boxes, for example, and their miniaturized CD successors give top billing to the opera title, the composer, and the conductor. The first two affirm the permanence of the work, certainly more so than the singers' names and record-company logos that dominate the labels of 78 rpm discs. But foregrounding the conductor has the opposite effect: when "Furtwängler" appears in the same font size as "Mozart," it is not clear whose *Don Giovanni* the box contains, nor how definitive it is. This particular convention persists in video packaging as well, in which composers regularly share the limelight with conductors and, more recently, stage directors. At the very least, opera "books" seem to have extraordinarily powerful editors.

Illustrations and explanatory texts go further, putting as much spin on the work as the performance does—sometimes more than one spin. In the 1950s and 1960s, the major labels and their affiliates frequently altered packaging to suit national markets, so the same performance might circulate with different cover images and booklets. Later, the multilingual booklets of CD sets could feature commentaries by different writers with different opinions, or libretto translations of varying age and viewpoint. Even when the contents are consistent, moreover, boxes and booklets say as much about evolving reputations as about original works. Together with opera guides, biographies, and other more traditional outlets for interpretation (and sometimes sharing the same authors), they mark successive generations of standards and taste.

In the case of *Don Giovanni*, the journey leads from picaresque adventure to dark cautionary tale, a transformation that consumes several decades and manifests most obviously in attitudes toward the title character. His likeness dominates box covers, which until the mid-1970s usually picture a dashing cavalier—Max Slevogt's portrait of Francisco D'Andrade updated for a new age. The original cover of EMI's 1959 recording shows him nonchalant in the familiar white doublet, tights, and sword, bathed in an equally familiar sunny light. A less familiar bouffant, together with a thin moustache and open collar, nudge his masculinity into the 1950s (fig. 5.1a). Though differing in style, several other covers from the 1950s and 1960s feature the same mixture of amused confidence and modernized Renaissance getup, whether in portraits or photographs.[12] The suave, charismatic Don Giovanni remains as popular a figure as he was in the first half of the twentieth century, and he makes a spectacular encore appearance on an EMI cover of 1974 as a many-hued charmer in psychedelic designs (fig. 5.1b).[13]

Otherwise the 1970s mark a change in both form and content, with original art giving way to reproductions and photographs, and nonchalance to less ingratiating attitudes. A new note of menace appears both on Deutsche Grammophon's recording of 1977 and Sony's movie soundtrack of the following year. In the first instance, the character's masculinity is updated again to take in the plunging necklines, tight pants, and smooth jaws of the disco era. More striking, though, is Sherrill Milnes's commanding pose, sword pointed groundward as if to demand submission (fig. 5.1c). Sony's cover features eighteenth-century garb, but Ruggero Raimondi's implacable gaze demands the same obedience as Milnes's sword, and it has become something of an icon in itself, surviving through many reissues and imitated in at least two other cover images.[14] Other releases use eighteenth-century paintings to evoke subtler forms of coercion, specifically the amorous scenes of Jean-Antoine Watteau and his successors. Though typically gauzy and impressionistic, the images have sharp edges of their own; in Watteau's *Le faux pas* of 1717, used on Harmonia Mundi's version of 2006, a woman tries to fend off the ardent embrace of a would-be lover.[15]

A few covers have no Don Giovanni, opting instead for empty stages, abstractions, or statues of the Commendatore, and still others summon the protagonist's sole moment of vulnerabilty: his punishment. The cover of Decca's 1955 recording has him being prodded into the flames by demons, his face frozen in fear and arms thrown up in desperation (fig. 5.1d). Reminiscent of comic books, the image constitutes a striking departure for its date, which may explain why the U.S. release (on Decca's affiliate, London) came instead with a photo of Cesare Siepi in cavalier mode. But the hero *in extremis* returns later, serving as the pathetic inverse of the brooding or sword-wielding predator. Most vivid is Eurodisc's own 1985 foray into pop culture, which features a stylized Alan Titus cowering before what looks like a Jedi light saber (fig. 5.1e). Sunny insolence forgotten, Don Giovanni becomes the antihero of a battle between good and evil.

EMI
MOZART
DON GIOVANNI
Eberhard Wächter
Joan Sutherland
Luigi Alva
Gottlob Frick
Elisabeth Schwarzkopf
Giuseppe Taddei
Piero Cappuccilli
Graziella Sciutti
Philharmonia Orchestra & Chorus
CARLO MARIA GIULINI

Mozart · Don Giovanni
Sherrill Milnes
Anna Tomowa-Sintow · Teresa Zylis-Gara · Edith Mathis
Peter Schreier · Walter Berry · John Macurdy · Dale Düsing
Wiener Philharmoniker
KARL BÖHM

EMI
EDINBURGH FESTIVAL PRODUCTION
MOZART
DON GIOVANNI
Roger Soyer · Geraint Evans · Antigone Sgourda
Heather Harper · Helen Donath · Luigi Alva
Alberto Rinaldi · Peter Lagger
The English Chamber Orchestra
Scottish Opera Chorus
Daniel Barenboim
conducting
FOUR DISCS · LIBRETTO IN ENGLISH

MOZART
DON GIOVANNI
DECCA
RECORDS
DONNA ANNA   SUZANNE DANCO
DONNA ELVIRA   LISA DELLA CASA
ZERLINA   HILDE GUEDEN
DON OTTAVIO   ANTON DERMOTA
DON GIOVANNI   CESARE SIEPI
LEPORELLO   FERNANDO CORENA
IL COMMENDATORE   KURT BOEHME
MASETTO   WALTER BERRY
with THE CHORUS OF THE VIENNA STATE OPERA
and THE VIENNA PHILHARMONIC ORCHESTRA
conducted by
JOSEF KRIPS

Wolfgang Amadeus Mozart · Rafael Kubelik
DON GIOVANNI
Alan Titus · Julia Varady · Rolando Panerai
Arleen Augér · Thomas Moser · Edith Mathis
Jan-Hendrik Rootering · Rainer Scholze
Chor und Symphonie-Orchester
des Bayerischen Rundfunks

FIGURE 5.1 LP album set covers, 1955–85

a. *Don Giovanni*, cond. Giulini 1959, EMI SMS 1009. Author's personal collection. Photograph by the author.

b. *Don Giovanni*, cond. Barenboim 1974, EMI/Angel SDL 3811. Author's personal collection. Photograph by the author.

c. *Don Giovanni*, cond. Böhm 1977, Deutsche Grammophon 2709 085. Author's personal collection. Photograph by the author.

d. *Don Giovanni*, cond. Krips 1955, Decca LXT 5103–5106. Author's personal collection. Photograph by the author.

e. *Don Giovanni*, cond. Kubelik 1985, Eurodisc 302 435–445. Author's personal collection. Photograph by the author.

"Wrongitude"

The texts inside the boxes tell a related story. Much like the examples discussed in chapters 3 and 4, translations of the libretto reflect shifting views of Da Ponte's sexual innuendos and especially his references to violence. For English-speaking consumers of its 1955 *Don Giovanni*, Philips recycled a late nineteenth-century text by the prolific translator Natalie MacFarren, who applies a sanitizing wash of respectability.[16] In the closing stanza of Zerlina's "Batti batti, o bel Masetto," she joins Friedrich Rochlitz and others in keeping Da Ponte's "night and day gaiety" firmly during the day. More startling, she tones down the graphic opening lines into something even more innocuous than "Chide me, chide me, dear Masetto": now, Zerlina sings "Canst thou see me, unforgiven, here in sorrow stand and languish?"[17] Leporello's act 1 reproach of Don Giovanni, "Bravo! Due imprese leggiadre; sforzar la figlia, ed ammazzar il padre," gets the same treatment. As noted in the introduction, even in recent years this line has given rise to considerable variation, with some video subtitlers rendering "sforzar" and "ammazzar" as "rape" and "murder" and others using the comparatively neutral "ravish" and "kill." MacFarren goes much further, having Leporello say, "Well done, the intention was pretty; seduce the daughter, and then dispatch the father!"[18] Like the sparkling aristocrats of the 1950s cover images, her Don Giovanni seems more naughty than criminal.

Translations change more quickly than album art, and the Philips label's use of MacFarren's text proves to be something of an anachronism. For HMV's 1936 recording from Glyndebourne, H. F. V. Little provided frank versions of the offending passages: Don Giovanni "rapes" and "murders," and Zerlina and Masetto are "happy and jolly day and night."[19] Among later translators such as William Murray, Peggy Cochrane, Lionel Salter, and Avril Bardoni, all of whose texts appear with multiple recordings, the principal departures are along the lines of "rape" versus "ravish."[20] Given the archaism of the latter term as a synonym for "rape" and its secondary meaning of "to overcome with emotion,"[21] its use indisputably softens Leporello's accusation. On the other hand, unlike MacFarren's "seduce" it retains the implication of Da Ponte's "sforzar," especially when paired with "kill" (Cochrane), or even "butcher" (Salter). Translations no longer uphold the "amorous Don" image.

It survives longer in the interpretive essays that accompany many recordings. From the late 1950s through the 1970s, these writings tend to echo nineteenth- and early twentieth-century claims about the protagonist, as well as the other characters and the drama as a whole. Karl Heinz Ruppel channels something of Victor Maurel's biological concept of the title character: "The class system . . . becomes insignificant when the entire social order is threatened by so vital and destructive a force, bound by no moral code, as the gay deceiver Don Giovanni. The principle for which he stands is the absolutely unbridled freedom of the life force."[22] To similar effect, the conductor Otto Klemperer quotes Kierkegaard's view that "its hero is the driving force behind the actions of the other characters. Don Giovanni's life is the vital core of their own existence."[23] Glen Sauls adopts a more political perspective. Riffing on the coincidence that *Don Giovanni* premiered in the same year that the United States held its constitutional convention, 1787, he describes the libertine as engaged in his own battle for independence: "He is damned because he flouted the conventions of continence and chastity, mocked the venerable institution of marriage, and espoused and practiced a philosophy of determined self-expression and individualism which was not in the interest of established authority." Larger than life and possessed of irresistible energies, the character deserves more admiration than critique. "It is in the sense that Don Giovanni defies convention and is a rebel against the proprieties that he engages our sympathies and cuts an attractive figure, perhaps even an heroic one."[24]

Other themes endure as well. Patrick Cairns "Spike" Hughes "defends" Don Ottavio by trafficking in stereotypes about weak men: "He is typical of the sort of man who almost invariably gives his doglike devotion to a woman like Donna Anna, whose choice causes her friends to shake their heads and wonder what on earth she can see in him."[25] He views Donna Anna with equal ambivalence, agreeing with E. T. A. Hoffmann that she falls willingly for Don Giovanni and grows angry only when she realizes he has stolen her honor (see chap. 9).[26] Once more the hero is exonerated of rape, and indeed the accompanying synopsis, also by Hughes, describes "Donna Anna . . . trying to unmask [Don Giovanni] after his attempt to *seduce* her" (my italics).[27] However full of life or rebellion, the libertine respects the line between persuasion and force. In the less polemical commentary of Irving Kolodin, a similar circumspection nonetheless helps redeem the entire opera: "Perhaps the subtlest appeal of *Don Giovanni* is its flattery of the mirror image that exists in the mind of most individuals of both sexes. Man is impulsive, virile, humorful, serious, honorable, and irresistible. Woman is unassailable, indulgent, susceptible, strong-willed, loyal yet desirable."[28] The most revealing words here are the ones Kolodin omits: "violent" for the men, and anything approaching "sensuous" or "erotic" for the women. Like so many of his contemporaries, he still wants listeners to imagine a swashbuckling costume drama.

This view would not survive the 1970s. The trend toward more threatening cover images tracked developments in contemporary performance, which in turn found their way into booklets associated with specific productions. In CBS's soundtrack album for the Losey film, an interview with the director puts even familiar themes in a new light. Losey argues that Donna Anna has taken Don Giovanni as a lover, but he also sees both as caught up in an epic transition from feudalism to revolution. A quotation from Antonio Gramsci prefaces both the film and the record booklet: "The old is dying and the new cannot be born; in this interregnum, a great variety of morbid symptoms appears."[29] Losey takes one of those symptoms to be Don Giovanni's profligacy, whose "bravura and panache" ultimately conceals a pathology: "Sex . . . at this kind of level is virtually masturbation and not a relationship. It seems to me clear that someone who runs from woman to woman like that—and tells—is trying to escape from something in himself."[30] Once inspired by the demonic, now the protagonist is merely tortured by inner demons. He fares no better in John Higgins's essay of 1984, which memorializes a Glyndebourne production directed by Peter Hall. Conceiving the opera as "a vision of the daemonic and a foretaste of the Romantic movement," Hall went so far as to set it in the early nineteenth century, the age of Hoffmann himself.[31] But his hero is no vessel of higher powers, as witnessed by his poor showing vis-à-vis a rehabilitated Don Ottavio: "He is the personification of rectitude while Giovanni is the personification of what can only be called wrongitude."[32] Defiant at every turn, he treats all around him rudely and tries to rape Zerlina. His faults are beginning to overwhelm his legend.

Continuing on the same lines, Stefan Kunze entitles his 1989 commentary "Comedy under the Sign of Catastrophe." The immediate catastrophe is the hero's fiery punishment, but more consequential are the effects of his misdeeds, which create a degree of suffering and trauma that the genre of social comedy cannot ameliorate, regardless of how many disguises and other comic conventions Da Ponte brings to bear. This is not merely a society being threatened, as in Ruppel's commentary; for Kunze, "The actions and relationships set in train by Don Giovanni undermine the preconditions for any kind of human community."[33] In the following year Paul Griffiths wondered whether the society represented in the opera was the only one being afflicted. His protagonist is also in flight, in this case from a past that he cannot or will not remember. When it finally catches up with him, the devastation is abstract (the "real punishment" is realizing that he is "a character in an oft-told tale"), but the aftermath is all too real.[34] Griffiths points to Donna Elvira, willing victim of so many humiliations, as symptomatic of a blindness that is shared by modern audiences: "She keeps coming back, even though she keeps getting nothing but mockery and pain, and it begins to seem that she too, like Giovanni, is unable to learn from the past. . . . She goes on putting herself through the same experience. As do we who go on asking this opera to tell and retell its story time and again."[35]

Works of Performance

Scholarship and criticism since 1990 have continued to dwell on the social costs of Don Giovanni's behavior in the opera, and of *Don Giovanni*'s survival in the repertory and curriculum (see esp. chaps. 6 and 9 below). Record booklets, however, have taken a different turn, focusing on historical background and performance. History always occupied a prominent place; nearly all of the writers quoted above also cover the premiere of Mozart's opera, and some discuss earlier adaptations of the Don Juan myth or the opera's reception by Hoffmann and others. Beyond context, such accounts offer reassurance that what the record set contains is the historical work, not merely an instantiation.[36] This in turn allows them be packaged with multiple recordings: in the mid-1990s, markedly different performances conducted by Charles Mackerras and Georg Solti were paired with versions of the same essay by H. C. Robbins Landon, a sober and thorough account that stops at 1788.[37]

Regarding performance, not every writer between the 1950s and the 1980s ignores it entirely. Though by no means an essay about Erich Leinsdorf's *Don Giovanni*, the Kolodin introduction quoted above does review the opera's transformation in the nineteenth and early twentieth centuries, and there is also a witty account of the recording process, couched as a series of letters between Mozart and his father.[38] The contingency of the accompanying discs is at least broached, if not exactly embraced. Two developments of the CD years helped tip the balance. One was the increasing importance of reissues, which necessarily lend new weight to interpretation. The most obvious case is EMI's 1986 release of Furtwängler's then thirty-two-year-old Salzburg Festival broadcast. Though not, strictly speaking, a reissue, that it appeared at all testifies to the reputation not of the piece so much as of the conductor and his collaborators. The packaging concurs: the cover pictures Furtwängler rather than Don Giovanni, and the essay, by the record critic Alan Blyth, deals with interpretation and broadcast rather than composition and premiere.[39] More recently the corporate successor of EMI, Warner Classics, accorded the same treatment to one of its own studio recordings, the 1959 version conducted by Carlo Maria Giulini. A longtime favorite among critics, it now comes in the same hardcover book format as Sony's Currentzis-led cycle of Da Ponte operas, the effect of which is to put Giulini on a pedestal, not to mention the soloists and the producer, Walter Legge, all of whom are memorialized in photos from the recording sessions as well as in a new essay by George Hall. All that remains of the original packaging is the Slevogt-like cover image and William Murray's translation; the original essay, Sauls's ruminations on Don Giovanni and the American Revolution, has disappeared. The consumer still gets an important historical text, but in a series dubbed "Legendary Opera Recordings," not "Legendary Operas."[40]

Historically informed recordings have also trained the focus on performance. CD booklets have been a prime venue for justifying the use of period instruments, unwritten ornamentation, and more, so much so that Richard Taruskin's critique of the movement targeted the booklets and their claims as much as (if not more than) the recordings themselves.[41] Often written by conductors, reflections on performance practice are as invested in the ideal work as any historical or critical essay; indeed, they often assert that the sound of the instruments, or the details of execution, or—in *Don Giovanni*—the sequence of numbers comes closer to Mozart's intentions than do the alternatives heard on other recordings. Paradoxically, the explanations often do as much to illustrate the particularity of the accompanying performance as its historical permanence. In an introduction to his 2006 recording for Harmonia Mundi, René Jacobs writes: "If I try to imagine how Hoffmann's *Don Giovanni* would sound, I hear strains of Wagner, or at most a completely distorted Mozart, played and sung much too heavily, with the wrong tempos, and, above all, much too uniform. All the subversion inherent to the *dramma giocoso* is condemned to oblivion by an interpretation of this kind."[42] The ensuing proof is particularly elaborate, with footnotes and a bibliography of sources in several languages. Yet its very thoroughness highlights the uniqueness of Jacobs's rendering, his own agency as much as Mozart's. His views could not possibly be printed with any other recording, and the level of detail, together with his admission of difference—other interpretations exist, however wrong—lays bare the ambiguities of the musical sources. No less than George Hall's encomium to Giulini et al., Jacobs's comments figure the performer as a creator, the performance as a work in its own right.

The performed work is highlighted by another element of historically informed sets as well. As noted above, about a third of the studio-produced *Don Giovanni*s present something other than the traditional composite version. The majority date from after 1990 and feature period instruments and practices, and their accompanying essays generally argue for the authenticity of either the Prague or the Vienna sequence. Surprisingly, however, only two examples offer one of those sequences by itself.[43] All the others include the alternative music as well, sometimes tacked on at the end, sometimes arranged so as to facilitate switching between one version and the other. In addition to restoring beloved arias, the intention is presumably to complete the historical record, to play everything Mozart ever wrote for *Don Giovanni*, right down to the long-neglected "Per queste tue manine." And yet nothing better illustrates the fragility of the authentic work: with the versions painstakingly distinguished in the track listings, listeners can experience for themselves the flexibility of Mozart's and Da Ponte's own theatrical practice. The exceptionally complete recording subverts its own completeness.

The Theater of the Mind

Before turning to musical contents, it is worth examining some further innovations of the complete opera format. One is the use of nonmusical sound effects. The producer quoted above, John Culshaw, who was the industry's leading voice on opera in the 1950s and 1960s, thought it his duty not simply to record scores but to make virtual dramas. To this end, his pioneering *Der Ring des Nibelungen* incorporates everything from the hammering of anvils to the collapse of Gibichung Hall. He and his collaborators sought sounds that were, in his words, "effective, dramatically speaking, yet which somehow did not disturb the flow and power of the music."[44] He further advocated the use of stereo, or more generally, "directional sound," to encourage a specifically theatrical manner of execution: "Stereo opera is, or should be, capable of suggesting much more than the simply physical effect of a singer's changing position from right to center, or right to left: It should give the singer freedom to move in such degree that, as on the operatic stage, the movement itself may well, to some extent, determine the inflection of whatever music phrase is being sung."[45] Culshaw and like-minded producers created quasi-stages in their recording halls, with microphones ranged across the front and grids drawn on the floor. Each square of the grids bore a number or other label, and singers could be directed from one to the next by means of a script.[46] Whether or not this influenced delivery, it certainly produces an illusion of motion comparable to that heard in stereo recordings of live stage performances, but without the stage or audience noise. In the "theater of the mind," as Culshaw called it, voices move in space amidst carefully chosen markers of stage action.[47]

His approach did not suit everyone, and *Don Giovanni* recordings evidence a spectrum of practices. Most like Culshaw's *Ring* is a *Don Giovanni* of 1959 produced by Eric Smith.[48] Sound effects occur frequently, from realistic sword-clanging at the beginning to a symbolic gong in the act 2 finale, ushering in the chorus of specters.[49] Directionality is even more common, especially in the conversational recitatives, during which the characters often move restlessly around the sonic field. Most recordings do less, using sound effects only when actions are mentioned by the characters or suggested in the music, notably the opening swordfight, Don Giovanni's tossing of coins to Leporello at the beginning of act 2, and his later beating of Masetto. Directional movement continues to enliven the recitatives,[50] but for the most part producers put clarity ahead of motion, exploiting the stereo field to separate the voices and the different sections of the orchestra. At the far end of the spectrum, a few versions eschew all movement and sound effects, as in Teldec's recording of 1988.[51] One of the first complete recordings to be made digitally, it would have impressed Culshaw for its noiseless background and distortion-free *fortissimos*, whose sheer impact may well compensate for the missing swords and gongs. Still, he would have missed the dynamic sound field and its illusions, the interplay between implied bodies along with voices.

That said, Culshaw also recognized the possibilities of what he called "purely *vocal* acting": "Whether on records or in the theater, it is invariably the mark of the greatest artists: and inasmuch as it is a question of phrasing, intonation, intensity or stress, it can be captured just as well monophonically as stereophonically."[52] Among record critics, singers' abilities to project drama in fact draws far more attention than producers' efforts to conjure stage action. As Nathan Broder put it in 1960: "Deprived of costume, gesture, movement, and all other visual props, [the singer] must create the character he is portraying entirely with his voice, coloring it in such a way as to reflect the different emotional states and mental attitudes he assumes in the course of the story."[53] Reviews often assess vocal acting, and Broder felt that one virtue of Deutsche Grammophon's 1958 *Don Giovanni* was the success with which its star, Dietrich Fischer-Dieskau, imitated the voice of its Leporello, Karl Christian Kohn, during the scenes in act 2 when the two characters disguise themselves as each other.[54] Delivery and tone color make the drama, not directional shifts or rustling clothes.

Though they are seldom mentioned by either producers or critics, complete audio recordings also rely on what I would call vocal sound effects. Just before they swap cloaks in act 2, Don Giovanni and Leporello have a long conversation, one of several sprinkled across the libretto. The libertine bribes Leporello to stay in his service, dismisses his plea to give up women, "justifies" his own libertinism, and lays out a plan for seducing Donna Elvira's maid. On the recording with Fischer-Dieskau and Kohn, produced by Otto Gerdes, the passage has no sound effects, not even a tinkling of coins at the beginning. It also has only one directional movement: Leporello going right to left in order to listen to the new conspiracy (1:06–1:14). On the other hand, there are several chuckles and vocal breaks suggestive of amusement, as at Don Giovanni's "Lasciar le donne! pazzo!" (Leave the women alone? Crazy fellow! [0:23–0:29]), or Leporello's "Non ho veduto mai naturale più vasto e più benigno" (I have never seen a greater or more kindly natural instinct [0:56–1:06]).[55] Together with changes in tone and aspirations, this turns the dialogue toward bromance, such that Don Giovanni sounds more high-spirited than nefarious, and Leporello more tickled than scandalized—for now. Vocal effects underwrite not only the characterizations, but the moral tone of the enactment. ♫

Talking Libertinism

The same passage illustrates a further, equally consequential development of the long-playing era. The men's conversation is one of many *secco* recitatives in *Don Giovanni*, the speechlike dialogues accompanied only by supporting harmonies. Between the 1950s and the present, they have evolved from a bridge between set pieces, rendered briskly and sometimes trimmed, into an arena of heightened expression in their own right. The change is not limited to recordings, of course; its causes include the advent of theater supertitles in the 1980s and the concurrent influx of directors from spoken theater and film who took fresh interest in the characters' verbal sparring.[56] Nonetheless, complete recordings provided another impetus, challenging performers to make recitative comprehensible and interesting in the absence of stage action.

Fischer-Dieskau and Kohn are fair representatives of standard practice around 1950. They race through their lines at the beginning of act 2, lingering only over significant turning points such as Don Giovanni's "Lasciar le donne!," which brings the first significant pause of any kind, and the ensuing rhapsody to women, which concludes with some modest sustain and vibrato. Otherwise the exchange is speedy and continuous, most strikingly at the beginning, where the bribe is offered and accepted without a hint of reflection on Leporello's part. Some earlier performers go still faster: in the live Met recording of 1942, Ezio Pinza and Alexander Kipnis finish the whole conversation in just over ninety seconds, ten seconds faster than Fischer-Dieskau and Kohn and with even fewer pauses or changes in vocal color.[57] ♫

A Don Giovanni of a later generation, the baritone Thomas Allen, described the style as singers "falling over their tongues demonstrating their skill in delivering recitatives more rapidly than machine-gun fire. It's the trap we all fall into, in the effort to show our Italian colleagues how diligently we've studied our text."[58] He gives a good example in his own first studio recording, the EMI-Glyndebourne release of 1984, where the act 2 recitative lasts only four seconds longer than that of Fischer-Dieskau and Kohn (1:46 vs. 1:42).[59] In 1990, however, in a recording made for Philips, the dialogue stretches to nearly two minutes, as Allen and Simone Alaimo dwell on the bribe and other twists and turns.[60] In an EMI version of the same year, William Shimell and Samuel Ramey tack on another fifteen seconds, taking still longer for the bribe and, in Shimmell's case, turning the libertine's self-justification into a lyrical arioso (0:50–1:09).[61] Subsequent versions pack in ever more nuance, culminating (for now) in Sony's Currentzis-led set, in which Vito Priante's Leporello voices anger and fatigue and exasperation all in the first few measures, and Dimitris Tiliakos's

Don Giovanni sings "Lasciar le donne!" three times instead of the two called for by Mozart (0:43–0:58).[62] Ninety seconds of repartee have become two and a half minutes of dramatic confrontation.

Changes in accompanimental style add nuance of their own. At the beginning of the LP era, most performances used a solo keyboard instrument

---

The Accompanists Speak

The early complete recording by Deutsche Grammophon (*Don Giovanni*, cond. Fricsay 1958) preserves the understated manner of recitative accompaniment typical throughout much of the twentieth century. Accompanying Dietrich Fischer-Dieskau and Karl Christian Kohn at the beginning of act 2, the uncredited harpsichordist relies primarily on resonant chords. Only key moments are highlighted, such as "Lasciar le donne!" or the defense of libertinism, which bring an abrupt cutoff and rolling arpeggios, respectively (0:21–0:35). The livelier style pioneered by John Constable (*Don Giovanni*, cond. Davis 1973) involves more substantive changes in texture along with pictorial effects. He switches manuals for Don Giovanni's lyrical moments, enfolding them with muted plucking, and he substitutes an ascending scale and arpeggio for what would otherwise be the sound effect of jingling coins. Over the decades accompanists have introduced several additional practices, many of which are on display in the most recent Sony release (*Don Giovanni*, cond. Currentzis 2015). The fortepianist Benoit Hartoin adds both a prelude and a postlude to the opening recitative of act 2, the prelude quoting the preceding duet and the postlude leading seamlessly into the following trio. He also supplies the coins (here a descending scale, 0:12–0:16) as well as a musical image for Don Giovanni's intended target, Donna Elvira's maid (similar to the opening of the trio, "Ah chi mi dice mai," 1:46–1:54). Meanwhile the cellist, Alexander Prozorov, furnishes atmospheric effects, sustaining the descent under Don Giovanni's rhapsodic "l'aria che spiro" (the air I breathe, 0:58–1:07) and scrubbing out a tremolo when the characters disagree at the end (2:20–2:24). Finally, Hartoin and Prozorov sit out Don Giovanni's self-justification, making way for the lutenist Israel Golani to provide a quieter background that also looks forward to "Deh vieni alla finestra"—for which the lute will return (1:12–1:27). Beyond dramatizing a single dialogue, the accompaniment clarifies its place in the unfolding story. ♫

playing block chords at each notated change of harmony, and the results can be very sparse. At the beginning of act 2 Mozart writes only five bass notes for the entire episode of the bribe, Leporello's plea, and Don Giovanni's dismissive response. If the keyboardist plays without sustain, as does the conductor Fritz Busch in 1936, the entries do little more than steer the harmony. More typically, they vary just enough to mark the principal turning points in the dialogue, such as "Lasciar le donne!" or the defense of libertinism. A similar style endures through the 1990s, providing occasional articulation within a general context of understatement. As early as 1973, however, the harpsichordist John Constable got more venturesome, introducing a profusion of rolls, cutoffs, and unnotated flourishes to illustrate the text. The critic Robert Anderson grumbled about "outrageous and unjustifiable capers," but within a few decades most players were doing as much if not more.[63] Among *Don Giovanni* recordings, the richest example to date is again Sony's version of 2015, in which an ensemble of fortepiano, cello, and lute provides a continuous stream of mimed actions and telegraphed sentiments. Like the singers with their copious pauses and vocal sound effects, the instrumentalists strengthen the scene's claim to independent significance, diminishing its musical distance from the surrounding ensembles. They also reinvent, or perhaps outdo, John Culshaw's notion of virtual staging. If directional movement and the odd sound effect imply a setting, an active accompaniment provides a running commentary, not unlike the strings of musical cues once used to interpret the streams of images in silent movies.

Recitative performance has had its own, in some cases substantial, role in the transformation of *Don Giovanni*. In the case of Fischer-Dieskau and Kohn, brisk delivery and a clipped accompaniment give further proof that the libertine and his sidekick are impetuous and insincere, reinforcing the chuckles, the general suggestion of high spirits, and even the cover image of the box, which pictures the title character as a dashing cavalier. By contrast, more deliberate performances often support the post-1970s revaluation of the opera, most obviously with respect to Leporello. When Simone Alaimo and Samuel Ramey pause before accepting the master's bribe, or when Tito Priante sighs and grunts and sounds irritated, they suggest that the sidekick is growing a conscience. The recitative converts Leporello from accomplice to accessory, someone who returns time and again to an experience he knows is perverse.

Don Giovanni's evolution is more complicated. Treating his recitatives as substantive utterances rather than hurried transitions foregrounds the casual effrontery and sexism of his words. On the other hand, sometimes the new deliberation allows for an intensity of feeling suggestive not of a murderous trickster so much as of a sincere, or maybe delusional, romantic. When Thomas Allen, William Shimell, and Dimitris Tiliakos sing the encomiums to women in the act 2 recitative, the character sounds like a seducer in full flood, complete with honeyed vocality and serenade-like backup—accompanying Shimell, the fortepianist Robert Kettelson actually quotes the act 2 Serenade, "Deh vieni alla finestra" (0:42–0:53). And yet Don Giovanni does not need to seduce anyone here, least of all Leporello, raising the possibility that he actually believes what he is singing. As the ebullient cavalier disappears—from cover images, commentaries, arias, and even recitatives—his replacement is not always a predator. He may also be a genuinely nice guy, at least in his own eyes, a romantic as convinced of his own sincerity as he was once contemptuous of sincerity itself.

CHAPTER SIX

Cruel Laughter

When *Don Giovanni* began losing its picaresque gleam, one element of the plot attracted special scrutiny. In 1983 Wye Jamison Allanbrook wrote, "There is one act which, although it has no place under the rubric of seduction and the announced themes of the opera, we are forced to witness over and over again, and which in the sheer cumulative weight of instances of inhumanity must have the most unsettling effect on the spectators—the constant savaging of Elvira."[1] The worst comes at the beginning of act 2, just after the recitative discussed in the previous chapter, where Don Giovanni's abandoned spouse is led first to believe he has repented and then to embrace a man who the audience knows is Leporello in disguise. As long ago as 1913 Edward Dent found this hard to stomach, and by 1989 Stefan Kunze was writing that the deception "undermine[s] the preconditions for any kind of human community."[2] Others would join Allanbrook in wondering specifically about its impact on the community of spectators. Donna Elvira "shames us for giving in to the entertainment and warns us that when we go home we shall be alone with ourselves," says Laurence Lipking, and Kristi Brown-Montesano asks whether the "golden boys Mozart and Da Ponte managed to 'out-Giovanni' the libertine himself, manipulating us with words and music into enjoying a sadistic display of power over a helpless woman."[3] Ostensibly comic but more plainly cruel, the scene questions the ethics of performing *Don Giovanni* at all.

To experience it on audio recording is to understand how much of the challenge is musical. Staging can heighten or ameliorate the cruelty, but our impressions depend equally on pacing, tone, and expressive detail, especially as they affect the trio for Donna Elvira and her tormentors, "Ah taci ingiusto core." Hailed by Dent as "perhaps the most beautiful number of the whole opera," it interrupts the men's prank with a moment of vintage Mozartian lyricism, an oasis of melody and delicately interwoven textures.[4] Performers of Dent's era tended to exaggerate the contrast, but later generations have effected a reversal, diminishing the trio's beauty in favor of its painful intrigue. Tracking along with commentators' revaluation of the scene, the shift tests the limits of musical reinvention: it is one thing to critique the action on paper, quite another to do so in performance, where the very quality for which Mozart is most often celebrated must somehow be stripped away. On record at least, the inhumanity of Donna Elvira's treatment can emerge only if the music sounds less than human.

All Too Human (Again)

Despite her treatment in the plot, or perhaps because of it, Donna Elvira fares better with commentators than any character save Don Giovanni himself. Hoffmann was unsympathetic ("the tall, gaunt Donna Elvira, with visible traces of a great but faded beauty"[5]), but most nineteenth-century accounts resemble Carl Adelmann's of 1888, whose title says it all: *Donna Elvira as Artistic Ideal*. "Elvira is a powerful being, made of flesh and blood, full of genuine inner life, with a nature that is fundamentally noble though not grand, and distant from every trace of the contemptible or ridiculous."[6] Should she appear ridiculous, the fault lies with "a deliberately degrading representation" or with the casting of inferior singers, a complaint echoed by Victor Maurel and others.[7] The terms change after 1900, but not the enthusiasm. Using a familiar phrase in a new context, Lipking speaks for others when he writes: "In fact she is all too human. Few if any of Mozart's characters contradict themselves so often, changing in a moment or a phrase from implacable resentment to abject surrender."[8] Her susceptibility distinguishes her from Donna Anna, whose fortitude earned a cooler reception (see chap. 9), and Donna Elvira's sensuality got a free pass even from the Victorians. Echoing defenses of Zerlina, Heinrich Bulthaupt declared that "an unearthly bond chains her to the dear man, but the earthly bond that ties Elvira to Don Juan is no less strong, and its power should be subjected not to a moral, but only to an aesthetic judgment."[9] Wayward but well-intentioned, she has all the imperfect beauty of a real person.

Or at least, of a treasured stereotype. Noting her similarities to earlier *donne abbandonate*, all of them likewise dreamed up by men, Lipking suggests another source for her popularity: "Hence men find her easy to read. The words that flow from the mouth of Mariana or Donna Elvira are just what a man might imagine a woman would say when deprived of his presence. She would, of course, want him back. He knows her text by heart because he wrote it."[10] No less attractive than her words, in this perspective, is the visual spectacle of her distress. Lipking cites a classic bit of voyeurism from Kierkegaard's *Either/Or* of 1843, in which the philosopher imagines Donna Elvira bursting from a remote cloister after a visit from Don Giovanni, veil torn and bosom heaving.[11] The art historian Heinrich Gustav Hotho beat him to it, having voiced a related fantasy some eight years earlier:

> I see Elvira before me, alive in her most beautiful budding out; she stands there with a tall, slender figure; her innermost soul gazes enraptured through the moist shining veil of her deep-blue eyes; the hair must be chestnut-brown, the forehead high; and around her tenderly swelling lips I perceive the signs of that unspeakable tenderness which flutters wistfully, as if in quiet pain, even in the anticipation of the greatest happiness.[12]

The difference from descriptions of Zerlina is telling, inasmuch as Donna Elvira's eroticism is understood in psychological as much as physical terms. Her sensuality comes from deep within, enduring though she knows that pain will follow. Her humanity is of a sort that satisfies a particularly invasive gaze.

As for her musical and dramatic representation, it has been in flux since the earliest days of the opera, thanks in the first instance to additions and cuts. Though apparent throughout, her internal conflicts come to a head in the accompanied recitative and aria added for Vienna, "In quali eccessi, o Numi! . . . Mi tradì quell'alma ingrata." A rare soliloquy in the opera (the only others are Leporello's "Notte e giorno faticar" and Don Ottavio's "Dalla sua pace," the latter also a Viennese addition), the number has her complaining of Don Giovanni's treachery, imagining his doom, and wondering at her own fatal attraction, each sentiment matched to a distinctive musical style. Without this scene, as in performances of the Prague version, she finishes the opera in seeming abjection, fooled by the disguises in act 2 and rejected outright at the final banquet. With it, as in the Vienna or the composite versions, she retains enough self-possession that her eventual declaration for convent life seems more like a decision, if a sad one, than a simple defeat.[13] Of similar consequence, though on a smaller scale,

A Dangling Recitative

At several points in the libretto, Da Ponte has one character react to another's aria in a solo recitative, then exit the stage without singing an aria of their own. Don Ottavio does so after both of Donna Anna's arias, and Donna Elvira after Leporello's "Madamina, il catalogo è questo." Apparently meant to cover scene changes as well as advance the plot, these "dangling" recitatives are sometimes omitted, especially in live performance where the singers must otherwise wait out the applause for their colleagues, then stay behind to add a potentially anticlimactic punctuation (in the Vienna and composite versions, Don Ottavio's brief statement after "Or sai chi l'onore" is usually repurposed as an introduction to his own "Dalla sua pace"). Studio recordings more often include the passages, especially after 1980. "In questa forma dunque," Donna Elvira's reaction to Leporello's aria, anticipates the later "In quali eccessi" in its mix of bafflement and rage over wandering harmonies. As the performance of recitative has grown more elaborate over the years (see chap. 5), some soloists have taken the opportunity to express real defiance. Maria Ewing, for example, crescendos through the character's opening realization of her deception, bursts out loudly at her call for vengeance, and remains forceful through the concluding "rabbia e dispetto" (rage and spite; the recording is *Don Giovanni*, cond. Haitink 1984). If only for a moment, the character appears determined to act rather than react, to take charge of her own fate. ♫

is a frequent cut in act 1. Between Leporello's "Madamina, il catalogo è questo" and Donna Elvira's flustered exit, the score and libretto call for a brief recitative in which she resolves to exact vengeance. Including the number strengthens her agency and deepens the poignancy of her subsequent travails, while omitting it heightens her seeming helplessness in the face of Don Giovanni's betrayal.

Ornaments and Sentiments

A famous Donna Elvira of the mid–twentieth century, Elisabeth Schwarzkopf (b. 1915), pointed to recitative performance itself as another variable of characterization, particularly the use of unwritten ornaments. As detailed in part I, singers born after about 1885 put new emphasis on the letter of

Mozart's scores, eliminating many of the high notes, turns, and other decorations added by their predecessors. The high notes have not returned (yet), but smaller ornaments have enjoyed a modest resurgence, notably the single-note dissonances known as appoggiaturas. Most often they are created by raising or lowering the first pitch of so-called feminine endings—that is, phrase endings in which two syllables occur on the same note, the first accented more strongly than the second. Such endings are ubiquitous among Mozart and his contemporaries, and in 1963 the conductor Charles Mackerras lamented that earlier generations "threw out the baby with the bath-water" when they banned their decoration. Summoning a range of early editions and singing tutors as proof, he argued that the appoggiatura "is not just an optional embellishment but . . . an essential part of the melodic style of all vocal music of the eighteenth and early nineteenth centuries, particularly the recitatives."[14] Coming to the same conclusion, the editors of the *Neue Mozart Ausgabe* recommended adding appoggiaturas to virtually every phrase of the composer's recitatives, in effect treating the ornament as a matter of diction more than expression. Though an improvised dissonance can add a shiver of emotion, the thousands of repeated-note endings in recitative cannot all be shivery: the strong-weak pattern of the appoggiatura more often functions as the singing equivalent of a comma or a period.

Not everyone was satisfied. Without criticizing the *Neue Mozart Ausgabe* directly, Will Crutchfield noted that many of the early sources call for more elaborate decorations than single-note dissonances, and also for ornamenting concerted numbers as well as recitatives, something the edition rarely suggests.[15] Conversely, Frederick Neumann accused the editors of going too far, ignoring melodic and expressive considerations that render ornamentation inappropriate. Though he, too, connected appoggiaturas to diction, he nonetheless declared that "words expressing firmness, finality, resolution, hatred, terror . . . should not be subject to the softening effect of the grace."[16] The evidence unearthed by Mackerras and Crutchfield suggests otherwise, but singers have tended to agree with Neumann. Schwarzkopf is a case in point. Her recordings include several versions of Donna Elvira's Viennese recitative, "In quali eccessi, o Numi!," a number that has no fewer than eleven repeated-note endings, distributed more or less evenly across the rage-filled opening and the sorrowful conclusion. Schwarzkopf changed her mind at least twice about how to execute them, adding no appoggiaturas in a 1947 excerpt recording or at the 1950 Salzburg Festival; then five, at Salzburg in 1954; then only one, in the 1959 complete studio recording conducted by Carlo Maria Giulini.[17] Some of the changes make comparatively little difference: in the weepy concluding phrases, the appoggiaturas of 1954 elaborate on strong-weak effects she had earlier achieved with dynamics. In the angrier opening, however, the appoggiaturas differentiate notes that Schwarzkopf elsewhere sings evenly, including on the colorful images "scia-gu-*ra*-to!" (wretch [0:20–0:24]), "sa-*et*-ta" (thunderbolt), and "*ca*-po" (head [0:41–0:51]). *Pace* Frederick Neumann, these words actually sound "harder" with ornaments than without, as the dissonances add new punch and echo the strong-weak endings in the accompanying orchestral motto. ♫

Nonetheless Schwarzkopf shared Neumann's unease, remarking of the 1954 version: "There's one very big portamento and then one very bad appoggiatura. . . . I eliminated a lot of them [appoggiaturas] because they weaken the character. . . . The musicologists are quite wrong there . . . they haven't been on stage!"[18] And indeed, in the 1959 recording she limits her decorations to a single word, the end of a sentence whose subject is emotional turmoil: "Che contrasto d'affetti in sen ti *na*-sce!" (What a conflict of affections is born in your breast! [1:00–1:14]).[19] The pitch in question is also approached from a third above, such that raising it creates a smooth scalewise descent. Only in such circumstances, where uncertainty is explicit and the musical effect of the dissonance minimal, would Schwarzkopf risk a "weakening" appoggiatura. Elsewhere she sticks to the score, associating the solidity of repeated notes with strength. (Ironically, she also complains that the 1959 characterization is "waspish" and "unsympathetic," but even this did not lead her to reconsider her practice.)[20]

The wariness felt by Schwarzkopf has only recently begun to fade. Other singers of her generation, such as Luise Helletsgruber (b. 1898) and Joan Hammond (b. 1912), likewise decorated few if any of the repeated-note endings in "In quali eccessi, o Numi"—a departure from predecessors such as Johanna Gadski or Astrid Lous (both b. 1872), who decorated most or all of them.[21] A few live recordings evidence more freedom, notably the 1942 Met rendition with Jarmila Novotná (b. 1907), who ornaments throughout even as Ezio Pinza does not. On studio recordings, however, sopranos would not be so venturesome until the 1990s, and even then they remain circumspect. Every "In quali eccessi" since 1990 has at least five appoggiaturas, but they nearly always occur in the concluding phrases, filling in thirds and embellishing the "softer" images of lamenting self-doubt. Other uses of the appoggiatura, whether to clarify the passage's diction or to "harden" its more aggressive words, have thus far been explored in only a handful of versions. Ornamentation has served largely to feminize Donna Elvira, emphasizing her capacity to yield rather than to defy.

Sincerity Meets Slapstick

Donna Elvira's deception at the beginning of act 2 raises another kind of performance issue. It bears some similarity to her first run-in with Don Giovanni and Leporello, in act 1: both scenes start with a bickering recitative for the men (prefaced in act 2 by a bickering duet, "Eh via buffone"), followed by a concerted number involving Donna Elvira, then a recitative in which she is tricked. But whereas the first encounter goes on to culminate in Leporello's "Madamina, il catalogo è questo," the second essentially sputters out. After the concerted number—the trio "Ah taci ingiusto core"—a recitative leads not to an epic showstopper but to the "lovers" running off to make way for Don Giovanni's "Deh vieni alla finestra." The ending is more a detour than a climax, leaving the shape of the scene open to debate. Especially ambiguous is the role of the trio, whose text pits Donna Elvira's sincerity against the men's deceit, but whose music spins webs of gorgeous melody for everyone. Is this a comic episode or a showstopper in its own right, placed awkwardly in the middle of the action? Does it fit into its surroundings or rise above them? Is its Donna Elvira "strong" or "weak"?

In the first complete *Don Giovanni* recorded under studio conditions, at Glyndebourne in 1936, the conductor Fritz Busch and his colleagues opted for showstopping. They lay some groundwork by drawing a strong distinction between the recitatives and the concerted numbers, not just in this scene but throughout. Typically for the mid-twentieth century, the dialogues unfold briskly to sparse accompaniments, with Busch playing simple block chords with minimal sustain (despite using a modern piano) and the singers marking only the most obvious turning points with pauses or changes in tone.[22] The concerted numbers go in the opposite direction, as attested by the reviewer Dyneley Hussey: "Every phrase, both of words and music, is given the fullest possible energy of dramatic and musical meaning. And what an orchestral performance Fritz Busch secured—never hurried, but always alive and beautifully moulded!"[23] Such deliberation and nuance would have stood out against the more dynamic style of the previous generation's most celebrated Mozartian, Bruno Walter (b. 1876). At the beginning of act 2, Busch (b. 1890) takes both of the concerted numbers about 10 bpm slower than Walter does in his live recordings, and the mood is dignified even in the men's initial duet. Under Walter's direction, Ezio Pinza

Men Arguing

Performances of the men's duet, "Eh via buffone," set the tone for the entire opening scene of act 2. The characters enter mid-argument, Leporello complaining that Don Giovanni has nearly gotten him killed (in the act 1 Finale), and Don Giovanni insisting it was all a joke. The duet ends there, followed by the recitative in which Leporello is bribed to stay and the next conquest planned out. In the early recording from Glyndebourne, Fritz Busch takes a moderate tempo of around 63 bpm, and John Brownlee and Salvatore Baccaloni project an equally moderate anger (*Don Giovanni*, cond. Busch 1936). Brownlee even strikes a conciliatory tone at the beginning, opening harshly, "Eh via buffone" (Come on, buffoon), but then relenting, "non mi seccar" (don't annoy me; Gronda, *Il Don Giovanni*, 45; Stivender and Webb, *"Don Giovanni,"* 339). Theirs is a gentlemen's disagreement, stylish and dignified, and it prefigures an elegantly poised trio. In the 1950s the duet is often angrier, thanks in some cases to faster tempos and dynamic movement around the stereo field (e.g., *Don Giovanni*, cond. Leinsdorf 1959, with Cesare Siepi and Fernando Corena; or *Don Giovanni*, cond. Fricsay 1958, with Dietrich Fischer-Dieskau and Karl Christian Kohn). The recording conducted by Carlo Maria Giulini relies more on accent and tone, reinforced by its dry, bright engineering (*Don Giovanni*, cond. Giulini 1959). After four scratchy chords from the normally silky Philharmonia Orchestra, Eberhard Wächter all but spits out the first line, emphasizing every downbeat: "Eh via buf-*fo*-ne, eh via buf-*fo*-ne, *non* mi sec-*car*." Giuseppe Taddei's Leporello is smoother but no less insistent, alternating clipped emphasis with heavy sarcasm. They prepare, in turn, a comparatively unsettled trio, the high point of a scene shot through with anxiety. In the rendition led by Roger Norrington, finally, metrical effects broaden the seeming scale of the disagreement (*Don Giovanni*, cond. Norrington 1992). Though no faster than Giulini's or even Busch's version, the performance moves in phrases rather than beats, as Andreas Schmidt and Gregory Yurisich stress every other downbeat with accents and "hardening" appoggiaturas. With so much emphasis on the two-measure groups, the time signature feels more like 3/8 (and occasionally 6/8) than the written 3/4, and periodic crescendos add further propulsion. This in turn prepares a scene in which everything feels harried and forced, including the 6/8 meters of the trio and of "Deh vieni alla finestra." ♫

and Alexander Kipnis stage a chaotic shouting match, but under Busch's, John Brownlee and Salvatore Baccaloni conduct a measured debate, calm and almost respectful. ♫

The trio exudes still more dignity. Setting out at a little over 100 bpm to the eighth note, Busch treats the 6/8 meter as lilting but not swinging, and the violins' sixteenth- and thirty-second-note passages are light and precise. Luise Helletsgruber executes the opening phrases with matching poise, adopting a full, richly wavering tone but also releasing at the strong-weak endings of each phrase (in this instance Mozart writes the appoggiaturas into the melody). The sole exception, an end-accent on "un em-*pio*" (impious man), suggests the anger running beneath the character's sadness, but the contrast is far less than it will be in recordings by Schwarzkopf and others. Helletsgruber's Donna Elvira remains lost in melancholy.

The mood is equally restrained in the ensuing appeals by Don Giovanni, the first of which is sung to Donna Elvira's opening melody recast in the dominant (1:04–1:51), and the second to a preview of the Serenade melody set in the remote key of C major (the flatted third, 2:04–2:41). Much as in the earlier duet, Brownlee sounds calm and serious, recalling Pinza's Serenade performances (and his own) in his rhythmic and tonal consistency. Every phrase is full of songlike simplicity, even the final cadence which Brownlee declines to heroize: rather than march down the scale to the tonic, he allows the seducer's voice to fade away ("pentito io *sono già*" [I am already repentant]).[24]

The temperature rises somewhat in the ensuing passage of reaction, which has Donna Elvira crying out in disbelief, Don Giovanni reiterating his sincerity, and Leporello muttering his own wondering reaction (2:41–3:10). But while the voices grow more intense, all three singers continue to avoid histrionics, most significantly Baccaloni, who—unlike many Leporellos—does not literally laugh when he sings, "Se seguitate, io rido" (If you continue, I will laugh). Likewise, in the concluding section, which reprises the opening in ensemble texture, Baccaloni and Helletsgruber execute the outer lines as a liquid duo, lending solemnity to their prayer for Donna Elvira's fortitude. Brownlee's inner part is lighter but not strikingly so, such that the libertine's smug words blend into a reigning legato further bolstered by lush orchestral playing ("Più fertile talento del mio, no, non si dà" [A more fertile talent than mine, no, is not granted]).[25] The performance renders the trio as Dent's "most beautiful number of the whole opera," more lyrical than comic by far.

Its beauty stands out all the more in light of the recitatives, which border on slapstick. Their quick delivery forestalls any too elaborate "virtual staging," but Baccaloni does enough vocal acting to paint Leporello as an avid and sometimes leering buffoon. In the recitative following the men's duet, he adds an "mm-mm!" to Don Giovanni's "Lasciar le donne!," stutters when asked if he knows Donna Elvira's maid ("io, no"), and whines like a child when told he must switch clothes with his master. Most suggestive, he begins laughing during Don Giovanni's justification of libertinism and continues throughout his own response (0:50–1:03). Though to some extent alarmed, his Leporello is also a cheerful admirer of the libertine's ways. Baccaloni sharpens the image in the recitative following the trio, in which the character is called upon to finish the sham seduction and admit that he enjoys it. Responding to Helletsgruber's wavery vibrato, which now sounds tremulous, the singer first adopts a decent approximation of Brownlee's own light and sober tone (the character is now disguised as Don Giovanni). As the conversation proceeds, however, he brings back the laughter and eventually sheers off into overtly parodic fervency in the lines leading up to "Carissima!" (Dearest! [0:57–1:15]). The faux rapture continues through the rest of the exchange, culminating in a literal enactment of his pledge, "Lo giuro, a questa mano che bacio" (I swear it by this hand, that I kiss [1:30–1:36]).[26] Having restrained himself in the trio, Leporello takes up his new role with knowing winks, all of them invisible (in this case inaudible) to a clueless Donna Elvira.

The mugging affects the character of the entire scene, as does the stark contrast between recitatives and trio. In and of themselves, giggles and leers might plausibly be taken to temper the situation's cruelty; precisely because of their dubious taste, they remind listeners just how absurd it all is and in so doing shift the target of satire from Donna Elvira toward the overwrought rhetoric of seduction. No one truly suffers, and indeed some recordings (though not Glyndebourne's) suggest that Donna Elvira is in on the joke, echoing Leporello with her own parodic fervency. Alternatively, Leporello's chortling might be taken to provoke uneasiness among listeners who find themselves laughing along, as they are heard to do on many live recordings. Does anyone really want to feel his glee vicariously? Reacting to the similarly madcap Catalogue Aria on the Glyndebourne recording, an early reviewer complained that "it displays so little respect for Donna Elvira that the idea of her listening to the bitter end is incredible."[27] In fact, the idea of the aristocratic Donna Elvira listening to anything Leporello says is incredible, yet so familiar that it may take a dose of real crudeness to point it out.

That said, it is all but impossible to hear the recitatives as absurd, much less consciousness-raising, when the trio is so serious. With a soprano as luminous as Helletsgruber's, men's voices as genial as Brownlee's and Baccaloni's, and an orchestra as lush as Glyndebourne's, the number simply cannot come off as farce. Its characters are feeling subjects, its Donna Elvira the "all too human" character of critical commentary, not a stand-up partner for Leporello. Beautifying the number may be an attempt to protect her—to claim, as critics did with regard to Zerlina, that Mozart's music allows Donna Elvira to transcend Da Ponte's prurience and low comedy. Heard in sequence with the recitatives, though, as a complete recording allows, the effect is if anything the reverse. The very elevation of the trio performance makes the surrounding deception seem all the more base, an occasion for audiences to reflect—at best—on the oft-remarked inhumanity of so much eighteenth–century humor.[28]

Comedy with an Edge

As captured on LP recordings of the 1950s, performers of the next generation play the beginning of act 2 similarly, as an episode of rough comedy interrupted by an oasis of true feeling. Details, however, begin to erode the sanctity of the trio. Several Leporellos giggle at "io rido," most noticeably Italo Tajo, who adds a second laugh earlier in the number;[29] and a number of Don Giovannis lapse into parody, especially in the C major passage where Cesare Siepi overdoes his accents and Dietrich Fischer-Dieskau frames the melody with a gasping in-breath and a shuddering final descent.[30] The character's ensuing threat to kill himself is another favorite for exaggeration, "m'uc-*ci*-do!" Thirty or forty years earlier such word-painting would have sounded "rhetorical," and indeed the earliest-born Don Giovanni to have recorded on LP, Mariano Stabile (b. 1888), seems in the C major melody to be harking back to his own high-drama "Deh vieni alla finestra" recording of 1926.[31] But for younger singers such as Siepi (b. 1923) or Fischer-Dieskau (b. 1925), both of whose Serenades are elegant and smooth, the point is to sound silly, to import a little satire from the recitatives into the trio. The number remains beautiful, but it is no longer a wholly protected space.

The most celebrated of the mid-century *Don Giovanni*s, EMI's set of 1959, illustrates some consequences for the drama. Singing the title character, Eberhard Wächter inspired some of the same gripes Ezio Pinza did in his first performances (see chap. 2). Spike Hughes called Wächter's forceful interpretation "coarse," and Nathan Broder wrote that "he gives the impression not of an aristocratic libertine, but of a petulant and rather foolish man."[32] Hughes also complained about a "quite extraordinary noisiness" in the recording, which he blames on the performance but may have had as much to do with the engineering.[33] While other versions of the 1950s were recorded in echoey churches and halls,[34] the EMI producer, Walter Legge, and his colleagues created a comparatively dry, bright sound in their Abbey Road studio in London. This brings the action closer and renders Wächter's intensity all the more apparent.

The conducting contributes further intensity. The sessions began with Legge's longtime associate Otto Klemperer (b. 1885) in charge, only to halt a few days later when he fell ill.[35] Recordings from those sessions have never been released, but Klemperer's aesthetic can be inferred from EMI's next *Don Giovanni*, made under his direction in 1966, as well as a radio broadcast he led in 1955. Both aim for spacious grandeur, especially the 1966 version, in which the conductor takes strikingly deliberate tempos throughout. His substitute in 1959, Giulini (b. 1914), brought a fleeter and punchier conception to the opera, and though it too raised some hackles, it earned praise as drama.[36] Surveying complete *Don Giovanni*s in 1992, *The Metropolitan Opera Guide to Recorded Opera* deemed this one "a revelatory theatrical experience. Not only is it brilliantly cast and stunningly conducted, it exposes in a breathlessly exciting performance the profundity of a legend often taken as merely melodramatic, amusing, and safely outdated."[37]

Hyperbole notwithstanding, seriousness of purpose is abundantly evident in the rendering of the beginning of act 2. Biting, aggressive disagreement spills over from the duet into the men's recitative, where a single directional movement at the beginning, bringing Leporello forward as if to accept the bribe from Don Giovanni's hands, sets the stage for a close-range confrontation. Wächter never relents, bringing feverish excitement even to the phrases justifying libertinism, and Giuseppe Taddei projects weariness and sarcasm but no amusement. By comparison with contemporaries such as Dietrich Fischer-Dieskau and Karl Christian Kohn, or Cesare Siepi and Fernando Corena—all still chuckling and full of wit— the pair seem almost dour. ♪

As the trio begins, Schwarzkopf puts a similar edge on Donna Elvira's melancholy. Amidst tremulous lyricism and silky orchestra playing, she anticipates her entrances and makes the key words "un *em-pio*" and "tra-di-*to-re*" fairly leap out of the texture with accents (0:26–0:32). Her character's anxiety appears to lie close to the surface. As if warned, Wächter's

Don Giovanni starts off sweetly, but soon enough his excitement gets the better of him, typified by a surprise acceleration across the measures in C major (from ca. 103 bpm to ca. 115 bpm [2:00–2:33]). At this speed the passage prefigures not only the melody of the later "Deh vieni alla finestra," but also its 6/8 pastoral swing, and Wächter draws the connection tighter by singing without audible parody. Apparently his character regards every seduction as deserving of the same effort, and the recitative following the trio leaves the impression that he has wholly overpowered the others. Donna Elvira is undone—Schwarzkopf borders on parody herself, with aspirations and sighs—and Taddei foregoes snickering in favor of sighs of his own. Both characters fall so far as to lose any perspective.

By playing the scene with minimal humor, the performance casts new doubt on its humanity. Already a sympathetic figure, Donna Elvira becomes even more so when her inner conflict is so palpable, her gullibility so painful. Nor does the trio make much of an oasis, what with Don Giovanni surging forward and Leporello, his reluctance elsewhere notwithstanding, giggling in the traditional spot. At moments, especially near the end, the lush intertwining of voices and orchestra may continue to afford refuge—this is still 1959, after all, still the Philharmonia Orchestra. But amidst so many signs of suffering and wicked intent, beauty itself begins to sound like a tool of deception. Rather than transcending cruelty, Mozartian lyricism becomes its agent, evidence that the opera is indeed far from "merely melodramatic, amusing, and safely outdated."

The Reckless Dash

Recordings of the next three decades mix silliness and seriousness in varying degrees, with the scales tipping toward the latter. Rarely exceeding Busch's tempo of approximately 100 bpm, the trio remains gravely expressive, its spell broken by giggles but less and less often by parody: much like the figure on box covers from the 1960s through the 1980s, the seducer trends from geniality toward aggression. The men continue to ham it up in the recitatives, but the laughter is countered by increasing signs of disapproval from Leporello. As I noted in chapter 5, the diminishing speed and increasing dramatization of recitative after 1970 makes room for the servant's exasperated sighs and resentful tones of voice, among other things. They can invite sympathy, even identification, voicing an ambivalence that listeners may wish to share.

Beginning around 1990, innovations in the concerted numbers add further edginess. Faster tempos bolster the urgency and alter the dramatic pacing, shortening the ensembles even as slower delivery lengthens the recitatives. The trio in particular does not loom so large when its pulse reaches 115 bpm or more, a tempo that shaves upward of thirty seconds off its traditional duration. Unprecedented emphasis on phrasing and on metric patterns, especially dance rhythms, injects a complementary sense of physical movement. If Wächter's sped-up C major melody takes on something of the Serenade's lilt, a performance with differentiated beats and separated phrases goes further, approaching an actual dance. Finally, the noisiness grows as well: thanks sometimes to period instruments and their less resonant tones, sometimes to performers minimizing legato and vibrato—and sometimes to both—the sound acquires a grittiness that Legge and his engineers would scarcely have dared.

Though strongly associated with the historically informed performance movement, such innovations spread well beyond the period-instrument *Don Giovanni*s led by Roger Norrington, John Eliot Gardiner, and others in the 1990s. Swiftness and noise suffuse "modern" orchestra versions as well: neither the Chamber Orchestra of Europe, with Claudio Abbado conducting, nor the Vienna Philharmonic itself, with Riccardo Muti, yields much ground to its period counterparts in this regard. Singing changes more haphazardly, but performance writ large undergoes as profound a transformation as it had in the 1920s and 1930s, and with a similar boost from technology. Just as electrical recording highlighted modernism's smooth tones and rhythmic synchronies, digital facilitated the new era's finely textured brightness. Against a noiseless background, every weighted beat and breathy lift, every passing ornament and crunching string attack emerges with perfect clarity.

It was in this context that the critic Edward Seckerson, writing in 1995, enthused about a dramatic renewal: "No doubt about it, period manners, period style, the primal push of the instruments, the cut and thrust of the rhythm and articulation have put the reckless dash of the narrative back into *Don Giovanni*."[38] The masculinist language is familiar, harking back to the descriptions of D'Andrade and Pinza quoted in chapter 3, and so is the equation of musical aggressiveness with manly energy. Less clear is whether recordings of the 1990s figure "the reckless dash" as a good thing. The 1992 set directed by Norrington exemplifies the new sound, featuring period instruments and well-articulated (and ornamented) playing and singing, all packaged in a dry sonic envelope. At the beginning of act 2, the most palpable consequence is an overarching momentum that reconfigures the connections between numbers. The duet is angrier still than

Wächter's and Taddei's, aided by a rhythmic swing that emphasizes the three- and six-beat groupings of the vocal phrases. The feud intensifies in the recitative, Andreas Schmidt remaining gruff even in the justification of libertinism and Gregory Yurisich signaling disapproval with a weary sigh. Rhythmic force then enfolds all three characters in the trio, where a pulse of 125 bpm and accents on the first and fourth beats at last foreground the written 6/8 meter. At this speed the opening unfurls in three grand phrases, and Lynne Dawson's Donna Elvira evinces a corresponding vigor: her tone is more troubled than melancholy, and her flashes of anger require less contrast than usual. The remainder of the number unfolds like comedy, all hasty mutterings and forthright appeals and flashing back-and-forths right to the end, where the phrases are as well articulated as at the opening and any shimmer is undermined by Schmidt's clearly audible asides. ♫

Comic does not necessarily mean funny. Schmidt also revives the tradition of parody, in this instance with a period-appropriate ornamental turn in the C major episode, among other effects. Conversely, Yurisich does not giggle, even in the lovers' recitative where he sings sotto voce and minimizes the satire. There is no smooching at "bacio." With Dawson affecting comparable restraint, Schmidt is left giving the only signs of enjoyment: a series of chuckles both before and after the others' encounter. His Don Giovanni plays the puppet master. The characterization is strengthened by the ensuing "Deh vieni alla finestra," which, despite being the goal of all the subterfuge, does not always measure up; in more than one recording, it trails the more elaborate trio like a wan shadow. With such a lively trio, however, the relationship changes: the quick 6/8 meter of the Serenade (nearly 170 bpm to the eighth note in this case, or ca. 54 to the dotted quarter) now seems like the next logical step past the moderate 6/8 of the trio (ca. 44 bpm to the dotted quarter), which itself echoes the 3/8 and 6/8 implications of the rhythmically enlivened duet. In retrospect, Don Giovanni appears to have had command of the music as well as the plot, bending a reluctant Leporello and an unusually self-possessed Donna Elvira to an arc of his own making. The dash is not reckless so much as heartless.

Several other, similarly paced versions of the 1990s leave a related impression. The Don Giovannis are sometimes friendlier than Schmidt's, but their *suavité* rings hollow against the relentlessly forward motion, and against the skepticism of Leporellos such as those played by Samuel Ramey, Ildebrando d'Arcangelo, or Alessandro Corbelli.[39] Nor does the newfound energy of Donna Elvira make her any less sympathetic; it is no easier to watch an invigorated romantic get fooled than a melancholic one. With the lyricism and laughter diminished, performances of the scene begin to feel like critiques, all the more so when the odds are stacked even higher against the victim. In a 2002 recording conducted by Bertrand de Billy, Kwangchul Youn's Don Giovanni is pushy and curt, and Maurizio Muraro's Leporello laughs before, during, and after the (very fast) trio while also panting audibly during the lovers' recitative. Heidi Brunner's Donna Elvira resists, but even the continuo accompaniment mocks her, embroidering her laments with "tears"—yet another use of appoggiaturas. Unmitigated by kindness or sincerity, the gags sound mean, their very exaggeration revealing their depravity. The sadism of the men, rather than the gullibility of the woman, is what sticks in the memory.

Not So Human

Or is it? Regardless how hard-edged, no performance will convert every listener into a critic, and on some post-1990 releases derived from live performances, audiences can still be heard laughing as the men play their trick. These are not the howls that greeted Alexander Kipnis's stuttering Leporello at the Met in 1942, and in recent years the laughter may well be nervous or disbelieving. But it may also be unregenerate: if Donna Elvira's humiliation is unsettling to some, to others it will always be risible.

In light of this, it may be telling that some recent recordings seem as determined to send up the absurdity of the scene as to probe or critique its emotional content. I argued above that Baccaloni's slapstick Leporello might have had this effect, if only the trio on the Glyndebourne recording did not seem so desperately sincere. Latter-day examples make an issue of their own artifice. The last chapter looked briefly at one case, the 2015 venture led by Teodor Currentzis. With an overbearing Don Giovanni, a disapproving Leporello, and a suffering Donna Elvira, the beginning of act 2 is in many ways familiar. A proliferation of added effects, however, foregrounds the rehearsed quality of their interaction. The special treatment of Don Giovanni's justification of libertinism—a switch from prosodic recitation to lyrical arioso in the voice, and from fortepiano to lute in the accompaniment (see. chap. 5)—prefigures an almost fussy pair of seductions. In the trio, both of Don Giovanni's appeals to Donna Elvira feature multiple changes of tone and dynamics as well as added filigree in the keyboard, including, in the C major episode, a lengthy near-quotation of the mandolin part of "Deh vieni alla finestra." The latter aria continues in the same vein, with fortepiano and lute harmonizing the mandolin solo and the voice adding frequent ornaments and tonal contrasts. Witty and showy, the musical fixings join the flowery language in marking the whole spectacle as over-the-top.

The sequence seems equally self-conscious in the 2006 rendering directed by René Jacobs. The concerted numbers are quick and the connection between their triple meters clear, yet still they make room for colorful orchestral effects and for many vocal ornaments and gasps, especially from Don Giovanni: Johannes Weisser executes his appoggiaturas with such swooning sensuality as to parody assumptions about "the softening effect of the grace." An active continuo group completes the picture, embroidering the dialogue with playful commentary. However reckless in its pacing, the recording shows the performers to be as clever and artful as the characters are conniving or suffering.

Reminiscent of the earliest audio recordings, such dramatizing of detail may prove to be a blip, or it may become so ingrained that it no longer highlights the work of performance. The Jacobs and Currentzis recordings may also be read differently: as simply funny, or as no less critical, in their own way, than any other post-1990 version. Yet to my ear they suggest a continuing desire to mitigate the ethical quandary posed by Donna Elvira's deception. Their interpretive excess helps to ward off emotional engagement, not unlike the masks and stylized poses used in the *Don Giovanni* productions of Achim Freyer, a disciple of Bertolt Brecht.[40] Emphasizing performativity creates distance, the famous Brechtian *Entfremdung* (alienation), whose evocation in this context may feel liberatory—or perhaps a little desperate. If the treatment of Donna Elvira cannot be rehabilitated, its inhumanity elevated over its humor, then the only recourse short of taking *Don Giovanni* off the stage may be to remind listeners that even the most human of Mozart's characters were never human to begin with.

Dancing in Time

In his manifesto for conductors, *The Composer's Advocate*, Erich Leinsdorf opens a chapter devoted to Mozart by declaring: "No matter how diverse their views on every other topic, most musicians agree that finding the right tempo is at least half the interpretation. Wagner went further, asserting that the right tempo *was* the interpretation."[1] Nikolaus Harnoncourt agrees, at least so far as *Don Giovanni* is concerned: "For me, the dramaturgy of tempo, which starts with the overture and which ends with the end of the piece, is, I think, the most important part of the architecture of the whole piece."[2] Both conductors have plenty of recommendations to make, and nothing occupies them more than the composer's operatic finales. Strung-together sections in different speeds and meters, full of action and uninterrupted by recitative, finales put tempo planning to the test like no other passages in Mozart.

The act 1 finale of *Don Giovanni* has drawn attention for other reasons, most famously its ballroom sequence calling for three orchestras to accompany three different dances—all at once. This extraordinary quodlibet sets up the most significant turning point in the plot, Don Giovanni's attempt to rape Zerlina, which exposes his crimes and sets him on the road to punishment. It also anchors a typically involved structure for an opera buffa finale, a sequence of twelve sections that eventually brings all the (living) characters onstage at once. Tempo alone does not determine the shape of their interactions, but it does play a substantial role, and its effects are enhanced on audio recordings, where the events must be imagined primarily from the music and sound effects. In plotting their approaches, Leinsdorf,

Harnoncourt, and their peers reveal some broadly differing tendencies, especially with respect to the relative weight given to section-by-section contrasts versus longer-range trajectories. Their choices reach to the heart of the drama, ceding more or less priority to the intricate struggles between all the characters, or to the storied energies of the protagonist. "The dramaturgy of tempo" writes its own chapter in the opera's transformation, this one about the competing claims of social drama versus myth.

Theorizing Tempo

Speculation about Mozart's tempos dates to the early nineteenth century, when the advent of the metronome led to publications suggesting numerical values to go along with the composer's written markings. Two examples concerning *Don Giovanni*, an 1822 vocal score published by Moritz Adolf Schlesinger (b. 1798) and an 1839 article by Gottfried Wilhelm Fink (b. 1783), make recommendations for every major tempo change in the score (table 7.1).[3] Schlesinger's numbers presumably reflect his own preferences as well as practices of his time, a generation removed from Mozart's final years. Fink's suggestions derive from a composer, Jan Tomašek (b. 1774), who heard the opera in Prague just four years after its premiere.[4] He was young at the time and did not share his memories until five decades later, but as a seasoned musician trained in the late eighteenth century, he has at least as much claim to authority as Schlesinger. Since their reprinting in the 1960s and 1970s, both sets of numbers have figured into modern discussions of tempo, forecasting and perhaps influencing important changes in practice. Speeds that would have seemed unusual in the early LP era, too fast or too slow, have since become mainstream (see especially the discussion below of the menuetto).

Modern writers have relied on two additional sources. One is the later eighteenth-century literature on musical performance, such as the wide-ranging manuals by Mozart's father, Leopold, a violinist and music director, or the composer and organist Daniel Gottlob Türk.[5] They offer general characterizations of the most common tempos and advice on determining specific speeds, which requires more than just a written marking: performers are urged to consider meter, the note value of the governing pulse (half note, quarter note, etc.), the character of the governing dance rhythm (if present), and the music's emotional qualities.[6] Similarly holistic thinking has been applied to a second source: autograph scores and other authentic copies of Mozart's works. The composer used dozens of markings, and

TABLE 7.1. Nineteenth-century metronome markings for act 1 finale

Measure no.	1	92	139	218	251	273	360	406	468	499	533	623
Marking	All. assai	Andante	Allegretto	Menuetto	Adagio	Allegro	Maestoso	Menuetto	All. assai	And. maes.	Allegro	Più stretto
Meter	¢	3/4	2/4	3/4	¢	6/8	2/4	3/4	¢	¢	¢	¢
Pulse	♩	♩	♩	♩	♩	♪	♩	♩	♩	♩	♩	♩
Schlesinger (1822)	112	80	112	92	96	126	58	100	108	80	126	152
Tomašek (1839)	100	84	120	96	104	126	80	96	116	84	88	no marking

considerable energy has gone into classifying his painstaking gradations, especially his many varieties of allegro and andante.[7] The assumption, again, is that a comprehensive view provides the best guide to specific examples. "In the final analysis," argues Harnoncourt, "we are forced to rely on comparative approaches for Mozart's music, i.e. we must compare all passages which have the same tempo marks. Some of these demand by their nature a certain tempo, which helps to reveal Mozart's system."[8] If he ever divined the whole system, Harnoncourt did not say what it was, but the challenge was taken up in an imposing study by Jean-Pierre Marty and another by Helmut Breidenstein.[9] Both assign every passage bearing a tempo indication to what Breidenstein calls "modules," which share meters, governing note values, and markings—the same criteria emphasized by the eighteenth-century writers. Both catalogue Mozart's entire oeuvre, and Breidenstein's inventory runs to 420 distinct modules.[10]

Though he stops short of proposing a system, Leinsdorf exemplifies a comparative approach when he discusses the act 2 finale of Mozart's *Marriage of Figaro*. "Figaro appears to the sound of a German dance [allegro, 3/8]. . . . The tempo of this piece can be easily ascertained, for Mozart wrote some forty-nine German dances." The subsequent passage, too, an andante in 2/4 in gavotte rhythm, can be interpreted in light of *Don Giovanni*'s "Batti batti, o bel Masetto" or Haydn's symphonic andantes.[11] His confidence belies some obvious uncertainties: neither Mozart's German dances nor Haydn's andantes come with composer-approved metronome markings, and they can serve as a guide only insofar as one believes that they "demand by their nature a certain tempo," to recall Harnoncourt's equally confident assertion. Even the most dedicated comparatists, however, regard similarity as only a starting point for interpretation, and Leinsdorf invokes other criteria as well. He writes that Susanna's entrance in the act 2 finale, for example, a 3/8 marked molto andante, must go quickly enough to telegraph her nervousness, and that the surrounding allegros need enough space between beats to accommodate syncopated rhythms in the strings.[12] Whatever tempos may be suggested by resemblances to other music, they must be modified according to local considerations, musical and expressive.

The order of tempos looms large as well, and the *Figaro* finale illustrates the divergent emphases mentioned above, on long-range trajectories versus successions of contrasting sections. Leinsdorf privileges the former: "This finale . . . consists of ten sections, which means that without any interruption the conductor has to build one long, continuous line over these 900-

odd measures."[13] His solution is to use four sections in 4/4 meter, conducted in two and marked with ever faster gradations of allegro, as the scaffolding for a grand acceleration that subsumes all the intervening episodes. Harnoncourt, on the other hand, focuses on the relationship of one section to the next, arguing that Mozart intends all the 4/4 sections to be performed

---

Proportional Tempos

In a study written in the mid-1960s, a decade and a half after he had conducted the very first LP recording of *Don Giovanni*, Hans Swarowsky claimed: "In classical-period music, there is no case in which changes are made to the course of the tempo within one and the same movement, other than on the basis of a clear relationship (proportia dupla, tripla, quadrupla, etc.)" (*Wahrung der Gestalt*, 62; on the date see p. 270). Accordingly, he proposed that nine of the act 1 finale's twelve sections employ the same pulse of 108 bpm, which is to govern half notes in the passages marked allegro, quarter notes in the andantes and the two menuettos, and eighth notes in the midpoint adagio. Two additional sections follow their own common pulse of 132 bpm (table 7.2). Around the same time, the musicologist Christoph Bitter suggested an even stricter scheme based on a pulse of 72 bpm, which he identifies with the menuettos and applies to all of the other sections and ultimately to every tempo in the opera (*Wandlungen in den Inszenierungsformen*, 33). In 1977 the conductor Michael Gielen offered yet another variation, this one with subtler gradations made possible by a wider range of mathematical ratios. Some parts of the finale run much as in Swarowsky or Bitter, with relatively constant pulses shifting between half note and quarter and eighth (e.g., andante–allegretto–menuetto), but at the end, a set of 3:4 relationships produces an acceleration reminiscent of Leinsdorf's *Figaro* finale (Gielen, "Über die Tempi," summarized in Wallnig, "Die Bewegung"). As far as performance is concerned, the only readily available *Don Giovanni*s conducted by Swarowsky and Gielen date from much earlier than their printed commentaries, and neither comes close to realizing their ideals (*Don Giovanni*, cond. Swarowsky 1950; cond. Gielen 1960). Most striking, neither observes a common pulse in the opening sections of the finale; as the note values shift, the beat varies by as much as 25 percent. If the conductors came later to abide by their proportions, the flow of the passage must have differed significantly.

TABLE 7.2. Proportional tempo plans for act 1 finale

Measure no.	1	92	139	218	251	273	360	406	468	499	533	623
Marking	All. assai	Andante	Allegretto	Menuetto	Adagio	Allegro	Maestoso	Menuetto	All. assai	And. maes.	Allegro	Più stretto
Meter	c	3/4	2/4	3/4	¢	6/8	2/4	3/4	c	c	c	c
Pulse	𝅗𝅥	♩	♩	♩	𝅗𝅥	♪	♩	♩	♩	♩	♩	♩
Swarowsky (1960s)	108	108	108	108	♪ = 108	132	♪ = 132	108	108	108	108	no marking
Gielen (1977)	92–96	88–92	84	84	♩ = 42	126	♩ = 63	84	126	76	100	132

in four, and at comparatively moderate speeds.[14] Rather than an intensification, they provide a foil against which the adjoining meters can suggest appropriate dramatic meanings. Harnoncourt evokes still another common principle when he recommends linking successive meters by way of common beats: "In the allegro 3/8 at the entrance of Figaro, the eighths should be just as fast as during the previous allegro [4/4], which makes the basic pulse—two eighths as compared with three eighths per beat—slower and thus more rustic; this is a country dance, after all."[15] Neither he nor Leinsdorf appears to regard continuity of beat as of paramount importance, but others have sought to calibrate entire sequences, including the act 1 finale of *Don Giovanni*, according to simple proportions such as 1:1 or 1:2.

Theory does not always translate into practice. The *Figaro*s of Leinsdorf and Harnoncourt reflect the broad conceptions they laid out in print: the former's 1959 version of the act 2 finale accelerates steadily through allegros that sound as if they are played in 2/2, while the latter's 1994 rendering remains steady in 4/4.[16] On the other hand, when Harnoncourt's Figaro enters in 3/8, the eighth notes do not carry over from the previous section as he imagined, but rather slow down by about 25 percent. In performance, the "country dance" requires more of a relaxation than he envisioned. Yet divergences in the recording studio do not diminish the value of the literature on tempo. Whatever the practical outcomes, the preoccupations of conductors and scholars—markings, meters, resemblances, continuities, long-range effects, evocations of dance—have clearly guided Mozart performance throughout the modern era. Observing their treatment helps clarify the singular dramas that divergent tempo plans can produce.

Measuring the Dance

In the act 1 finale of *Don Giovanni*, the best place to begin is with the menuetto. It stands out for several reasons, beginning with its status as the most ubiquitous of the eighteenth-century dances: Mozart alone wrote hundreds, providing ample grist for comparison. It also underpins the fateful ballroom sequence, occupying the first of the three orchestras while the second and third drape contrasting dances over its beats, a duple-meter contradance and a triple-meter *Teutsche*—the German dance referred to by Leinsdorf in his discussion of *Figaro*. Together they set the abduction of Zerlina to what a filmmaker would call source or diegetic music,[17] sound that emanates from within the world of the characters. The distinction proves critical in audio recordings, which must somehow suggest the visual spectacle of characters dancing and orchestras playing in different places onstage. Equally important, both the menuetto and the contradance appear earlier in the finale (see tables 7.1 or 7.2 above, or table 7.3 below). The contradance sounds offstage at the beginning of the third section (allegretto), prompting Don Giovanni, Zerlina, and Masetto to depart for the ball, and the menuetto does the same in the fourth section, accompanying the maskers' negotiation of their own entrance to the party (Donna Elvira, Donna Anna, and Don Ottavio, disguised in order to catch Don Giovanni in the act). Nearly all recordings observe the same tempo in the first appearance of the menuetto as in the later ballroom sequence, and some adopt its quarter-note pulse for the first hearing of the contradance as well. As a result, a significant proportion of the finale may unfold to the same or a very similar beat.

Over the years, the speed of that beat has covered an extraordinary range, from 70 to 110 bpm with outliers at both extremes (fig. 7.1). The effects on the drama are equally extraordinary, though some qualifications are in order. While written tempo plans typically give only one metronome marking for each section of the finale, performances are rarely so simple. As we shall see, some sections are subdivided by what amount to unmarked tempo changes, readily audible shifts highlighting one or another distinctive turn of events. Such is not the case in the ballroom sequence, but here as elsewhere it is not uncommon for the tempo to wax and wane a little, usually by 3–5 bpm. This may pass unnoticed by those not listening with metronomes, but there can be subtle effects nonetheless: in some versions the last part of the dance scene speeds up slightly, tracking the growing dramatic tension,[17] whereas in others the same passage slows down slightly, accommodating the increased complexity of texture brought on by the entry of the third orchestra.[18] Unitary tempo markings are useful for planning and comparison, but they are best read as connoting the kind of narrow ranges indicated by the conductor Michael Gielen in his finale plan (see table 7.2).

A second qualification concerns the acceleration of the menuetto pulse over the last nine decades, which is real but also complicated. The trend lines are clear both among studio-produced audio recordings of the complete opera (fig. 7.1a) and across all the studio-produced audio recordings, archival live audio recordings, and video recordings I have studied (fig. 7.1b). Much as we saw in the last chapter, 1990 constitutes something of a dividing line, with only about 20 percent of recordings from before that date but nearly 70 percent of those after it playing the menuetto at around 90 bpm or faster (the midpoint of the overall range of 70–110 bpm). At the

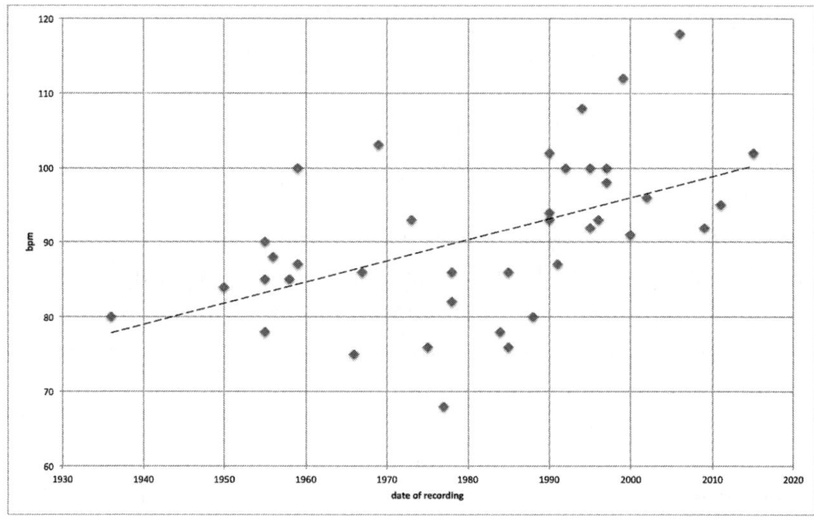

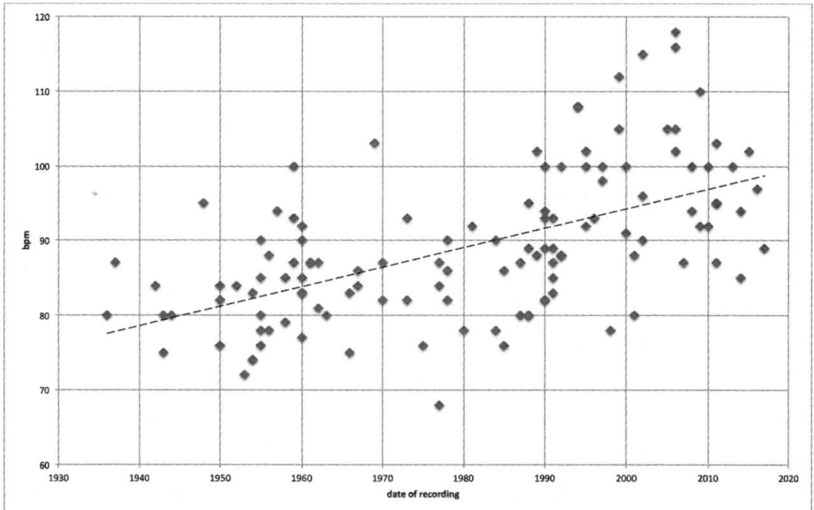

FIGURE 7.1 Menuetto tempos, 1936–2017
a. Studio-produced audio recordings
b. Studio-produced audio recordings, archival audio recordings, video recordings

same time, the progression is anything but uniform, and most individual decades encompass a range of at least 20 bpm in and of themselves. Generational and stylistic differences play some role in this: among recordings released between 2000 and 2010, for example, the birth years of the conductors range from 1929 to 1975, and the sonic ideals from modern orchestra lush to historically informed crisp. But individual choices are important as well—not least, in this case, those rooted in researches on tempo. The oldest conductor in the 2000–2010 group is Harnoncourt, a proponent of historically informed interpretation throughout his career. Yet he is also the source of the slowest menuetto tempo of the decade, about 80 bpm—a result of his particular reading of the historical record. The conductor regarded the *Don Giovanni* menuetto as the final stage in a long deceleration, a hundred years in which a dance once called "rapid" and "merry" became associated instead with adjectives such as "noble" and "moderate." Equally important, the same years witnessed the rise of quicker triple-time dances, predecessors of the waltz, with rapid steps and one-to-a-measure accentuation. The German dance is an example.[19] Faced with a superimposition of the two trends, Harnoncourt took the menuetto at a moderate pace and left the *Teutsche* to represent the newer, faster style, with the result that both his audio recording of 1988 and his video recording of 2001 echo the tempo preferences of still earlier generations, conductors such as Fritz Busch or Wilhelm Furtwängler.

By the same token, the generally faster menuettos of other post-1990 versions reflect not only the general trend toward fleetness and grit discussed in the last chapter, but an alternative reading of the evidence. While also recognizing a meeting of new and old in *Don Giovanni*, William Malloch argued in 1985 that, by the late eighteenth century, all minuets had quickened. By way of evidence, he cites surviving mechanical instruments along with early nineteenth-century metronome markings, among them Jan Tomašek's memory of hearing the *Don Giovanni* ballroom sequence played at 96 bpm (see table 7.1).[20] Malloch's aim was to encourage faster performances of symphony minuets, but the idea that all late eighteenth-century minuets are really proto-waltzes clearly affected *Don Giovanni* as well, both period- and modern-instrument renditions. From Roger Norrington to James Levine, John Eliot Gardiner to Riccardo Muti, the menuetto becomes brisker after 1990, and the German dance—as Malloch had put it—"heady and athletic."[21]

Speed alone has some impact on the character of the drama. Speaking very broadly, as its tempo has sped up the menuetto has rendered the entire dance scene more urgent and chaotic, supporting, in some cases, the increasing tendency to represent Don Giovanni as more rapist than seducer. Especially in performances where the surrounding tempos also quicken, the whole second half of the finale may feel angrier, or more comedic in the same humorless way that I discussed with regard to the opening of act 2 (see chap. 6). Beyond speed, accentuation and phrasing may also exert an influence. Speaking broadly again, over time the three beats of the menuetto have become more differentiated. Recordings of the 1950s tend to spread the emphasis equally, suggesting evenly weighted steps and casting the "old" dance as a kind of neutral foundation for the musical and dramatic events to come. Three decades later it is more likely that the menuetto downbeats will be strong and the offbeats weaker, and the meter articulated so as to emphasize the two-measure patterns of the melody. Inspired by eighteenth-century accounts, such performances imply a more independent choreography, a menuetto that drives the action rather than merely underscores it. ♩

At the opposite poles of speed and accent, the title character may seem to pursue different strategies, on the one hand harrying everyone forward, on the other using the dance to bewitch them. The 1959 version conducted by Giulini offers a good example of the latter, predicating the entire scene on an even, almost placid menuetto that seems to trap all the characters except Don Giovanni in a kind of spell. At a middle-of-the-road clip of around 87 bpm, the instrumentalists of orchestra 1 only occasionally stress the downbeats of the dance or its punctuating dotted rhythms.[22] All the variety comes instead from the singers (whose "vocal acting" is especially vivid), and from the subsequent entries of orchestras 2 and 3. Mozart introduces them during successive iterations of the menuetto's second strain, where first the contradance group and then the German dance band play open strings and trills to simulate tuning and warming up. They strike up the dances proper as the menuetto starts over, and in 1959 they figure prominently in the audio mix, their sounds a little louder than that of the menuetto and separated into right (contradance, beginning at 1:13) and left (German dance, 1:45) stereo channels. The original dance recedes further into the background, maintaining a constant line against the conflicting downbeats of the contradance and the triplet subdivisions of the German dance.

The underlying restraint of the menuetto has consequences. As its last iteration begins, with the other orchestras going as well, Don Giovanni sings a final "Vieni" to Zerlina, and she in turn cries foul: "Oh numi! son tradita!" (Oh gods! I am betrayed!). Subsequently, according to the stage directions, "ballando conduce Zerlina presso una porta e la fa entrare quasi per forza" (dancing, he leads Zerlina to a door and makes her enter almost by force). They are observed by Leporello, who follows them off after a hasty aside. Then come five crucial measures, part of the second strain of the menuetto, that may serve as either a further intensification or a moment of suspense (2:08–2:12 in the Giulini recording). The polymetric confusion continues, but the three maskers, who recognize what is happening, react with a quasi-proverb sung to the original rhythm of the menuetto: "L'iniquo da se stesso nel laccio se ne va" (the wicked man, on his own initiative, goes into the snare).[23] Their reassertion of aristocratic decorum, a matter of formal language and rhythmic poise, stabilizes the entire texture in some versions, making even the 3/8 of the *Teutsche* sound like a triplet accompaniment to the menuetto's 3/4—as if the whole edifice were really to be heard in a complex but unified 9/8. The stability may then endure through the ensuing measure and a half of vocal silence, the longest break since the beginning of the scene, which separates the maskers' cadence from Zerlina's offstage cry for help.

Giulini's orchestras do not quite resolve into 9/8, but the reaction of the maskers is weighty enough to restore some politesse, a sense of order that only the cry can disrupt. In retrospect, the menuetto's evenness and moderation seem like signs of passivity rather than neutrality, coding a respect for convention that Don Giovanni turns to his advantage. The Zerlina of the 1959 cast, Graziella Sciutti, increases the suspense by waiting the full measure and a half before crying out. Many performances fill in the gap with an unpitched scream, inspired perhaps by the stage directions for the following scene ("Si sente il grido e lo strepito dalla porte opposta" [from the opposite door are heard screams and clamor]), or perhaps by a disturbing anecdote from the rehearsals for the 1787 premiere, during which Mozart supposedly pinched his soloist to elicit an appropriately terrorized outburst.[24] Whatever its source, the extra scream obscures the last few seconds of dancing, turning the attention from onstage waiting to offstage struggle. In its absence, the lockstep of the menuetto persists a little longer, feeling all the more like a spell.

In differing degrees, performances like that on the 1959 recording suggest a similar connection between stately dignity and inaction. It is as if Don Ottavio's stubborn incredulity about the libertine's crimes (see chap. 3) had worked its way into the characters' very bodies, making them hesitant to move even when they catch Don Giovanni *in flagrante*. Running about 10 bpm slower than the Giulini recording, the versions led by Busch in 1936 and by Josef Krips in 1955 summon even greater degrees of inertia. Recorded in mono and thus necessarily more difficult to distinguish, orchestras 2 and 3 never really disrupt the solemn tread of the menuetto, and Zerlina enters—even with the scream, in 1936—as if to break the others out of a trance. Stereo recordings, such as those led by Otto Klemperer in 1966 and Karl Böhm in the following year, more adequately capture the rhythmic confusion, yet both of these versions still give precedence to order, highlighting the steady certainties of a dance that dips almost to 70 bpm, in Klemperer's reading. Two decades later Harnoncourt's calm approach produces a more lightly articulated but still quiet and slow-moving scene, a dreamscape for everyone apart from the perpetrator and his victim.

A 1995 set conducted by Charles Mackerras shows the difference a more dynamic menuetto can make. Orchestra 1 sets off at approximately 100 bpm, with all the instruments save the woodwinds (which have slurs) playing with light separation. Each two-measure group grows toward the second downbeat, injecting the rhythm with a strong forward lean and lending a degree of urgency to the scene-setting remarks at the beginning. As the interaction develops, two elements push the tension higher. One is the relatively forward mixing of all three orchestras, which typifies CD-era recordings and helps level the balance between instruments and voices. Both the rhythmic propulsion of the menuetto and the later confusion of the polyrhythms become more urgent when they are easier to hear. The use of solo first violins in orchestras 2 and 3 compounds the effect. In their efforts to distinguish the contradance and German dance, performers and producers across the years explore several means beyond stereo separation, including smaller ensembles, whose thinner sound is quite noticeable on record, and contrasting articulations. Regarding the latter, Mackerras's soloists go in opposite directions, the contradance leader playing with less separation than the minuet band, the German-dance leader with more. Meanwhile the solo sound itself cuts through the ensemble and allows each player to tweak the prevailing texture with ornaments and rhythmic anticipations. Their characterlike presence brings further animation to the critical final measures, the maskers' proverb and Zerlina's cry, during which the flourishes and syncopations of the German-dance fiddler remain especially prominent. Initiated with a bloodcurdling scream, Nuccia Focile's entrance feels in this context neither premature nor disruptive, but climactic. There is no spell to be shattered here, but an intensifying drama to be taken to the next level.

In differing combinations, quick tempos, propulsive beat patterns, instrument-heavy mixes, and solo violins tilt many other ballroom sequences from calculated suspense toward agitated struggle. At Mackerras's speed of circa 100 bpm, which is heard on several recordings from the last thirty years, the governing menuetto rhythm never really establishes a strong sense of decorum; on the contrary, the mood is restless from the beginning, and it only becomes more so with the fragmented vocal entries and, eventually, a necessarily frantic German dance.[25] In the few instances where the menuetto pulse is still faster, as in the recordings led by Daniel Harding (ca. 112 bpm) and René Jacobs (ca. 118 bpm), the whole sequence becomes frantic, flying by in a chaotic blur of stinging repartee and hyperactive fiddling. Catastrophe no longer sneaks up on the characters, concealed by bodily proprieties. It rushes at them faster than anyone can comprehend, save perhaps its impatient instigator.

Going to the Ball

The dancing takes place in the second half of the finale, to which we shall return. First, however, the guests must get to the ball, beginning with Zerlina, Masetto, and Don Giovanni, who spar with one another both before and after yet another attempted seduction; and continuing with the three maskers, who fret about their undertaking and pray for its success (table 7.3). This all happens outdoors in a garden, and the tempo markings and meters suggest a twofold deceleration, from fast to moderate (allegro assai to andante) and from medium-fast to slow (allegretto to menuetto to adagio). The trajectory looks even simpler if the third and fourth markings, allegretto and menuetto, inspire the same pulse. As noted above, the allegretto begins with a fragment of the contradance, leading some conductors to maintain a common quarter-note beat through both sections. Since the late 1980s, performances of otherwise differing aesthetics—for example, those led by Harnoncourt, Daniel Barenboim, Gardiner, Jacobs, Neville Marriner, and Arnold Östman—have traversed the whole sequence in four

TABLE 7.3. Act I finale, first half

TEMPO MARKING	METER	CHARACTERS	EVENTS
Allegro assai	¢	Z, M	Argument
		Z, M, DG, chorus	DG enters
Andante	3/4	Z, M, DG	DG attempts seduction
Allegretto	2/4	Z, M, DG	Contradance begins; Z, M, DG exit
		DA, DE, DO	Maskers enter
Menuetto	3/4	DA, DE, DO, L, DG	Menuetto begins; maskers converse with L, DG
Adagio	¢	DA, DE, DO	Maskers pray and exit

or five big steps, each one with a steady beat knitting together entrances, exits, and verbal back-and-forth into a focused and continuous narrative.

Not every performance aims for such continuity. The majority of conductors from earlier generations, and even some of their successors, complicate matters by introducing one or more additional, unmarked tempo changes. The first splits the opening allegro assai in two by means of an acceleration at Don Giovanni's entrance. Smartening up his arrival, which features a chorus of servants with trumpets and drums, the leap forward also summons the old critical commonplaces about demonic and biological energies, and it is presumably no accident that the conductors in question— Bruno Walter (b. 1876), Otto Klemperer (b. 1885), and Karl Elmendorff (b. 1891)—received their training in an era when the libertine's vitality was a given. They may well have thought it natural, even necessary, to introduce him with a rhythmic jolt.

Modifications to the allegretto section have endured longer, beginning with a kind of exclamation point midway through. The opening contradance quotation, played offstage by the same orchestra 1 that later strikes up the menuetto, lasts only eight measures. Its rhythm is the signal for Don Giovanni, Masetto, and Zerlina to leave the garden for the ballroom, but first they spend another twenty measures expressing anticipation, rejoined by the main orchestra, which adds thrumming repetitions in the bass and whirling sixteenths in the violins. Driven by a crescendo, this passage builds to a cadence on F (the key of the preceding andante), and in many recordings it also speeds up, right away or during the course of the crescendo. The moment of anticipation becomes a stretto, running 10–20 percent quicker than the contradance and providing the first dramatic episode of the finale with an emphatic conclusion. ♫

A still more frequent change follows when flourishes in the violins lead to D minor, fresh sixteenth-note figurations, and the entrance of the maskers. At this point the majority of performances slow down: conductors who have accelerated return to the original pace or slower, and those who have not introduce new tempos. Studio recordings of the 1950s evidence the various possibilities, as shown in figure 7.2: Max Rudolf remains steady through the F major cadence and then slows down at the arrival of D minor; Giulini accelerates and returns to roughly the same speed; Ferenc Fricsay accelerates and then drops to a slower tempo. (Readers are reminded that the speed of a given section or subsection may be subject to modest variation, as discussed above. The graphs are designed to illustrate long-range trajectories, not the moment-to-moment flexibility that characterizes most

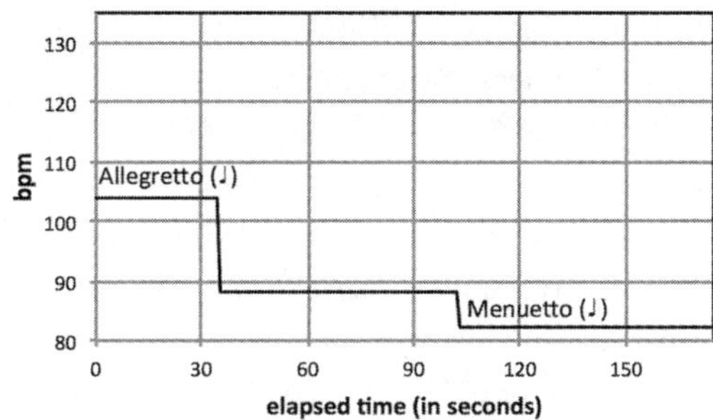

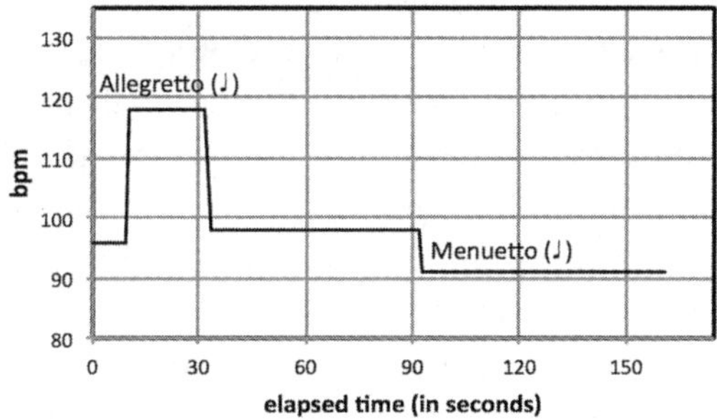

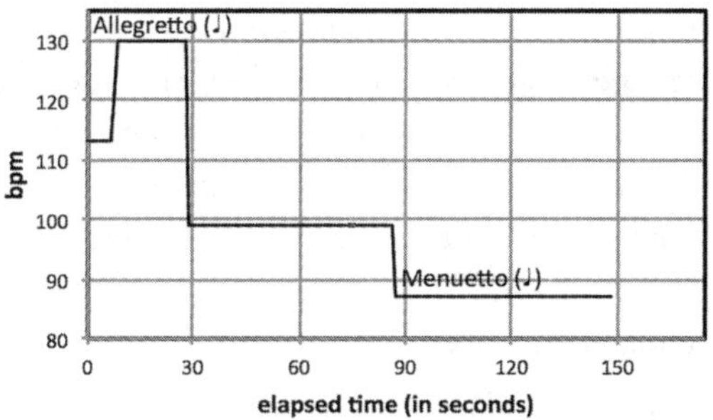

FIGURE 7.2 Act I finale: Allegretto-Menuetto tempos
a. *Don Giovanni*, cond. Rudolf 1955 (begins at 3:36)
b. *Don Giovanni*, cond. Giulini 1959 (begins at 1:57)
c. *Don Giovanni*, cond. Fricsay 1958 (begins at 2:05)

performances.) Heard frequently even in the most recent recordings, the deceleration at D minor in particular intervenes in both the music and the drama.[26] Key change notwithstanding, Mozart's music for the maskers carries on in the spirit of the preceding, with an active accompaniment framing agitated dialogue. If anything, the new staccato markings and neighbor-note patterns in the orchestra, together with the teetering descents and chromatic neighboring figures of Donna Anna's entrance, communicate more urgency rather than less. By opting for deliberation, performers play instead on the fateful associations of the key—D minor, heard in the overture and the statue scene—and on the potential for Donna Anna's music, including its chromaticism, to suggest something other than the fear she expresses in her words. At a slower pace she can assert some dignity amidst the fracas, such that she and her companions do not appear to join the plot so much as to start a new one. Their arrival transforms the garden into a place for high-flown reflection and lament.

It may also forecast a momentary transcendence of intrigue. At the end of the D minor passage, which is also the end of the allegretto, the offstage orchestra 1 strikes up the first menuetto, and Leporello engages the maskers in dialogue. When not taken at the same quarter-note pulse as the allegretto, the menuetto nearly always goes slower, making another stage of deceleration for those recordings that have already slowed down at the transition to D minor (as in all three recordings in fig. 7.2). Then comes the biggest deceleration of all, to adagio, where recordings of all eras most often observe a roughly 1:1 proportion between the quarter notes of the old section and the eighth notes of the new. Dating as far back as Schlesinger (see table 7.1), the proportional relationship may be open to dispute inasmuch as Mozart notates the adagio in two beats per measure (¢), which would seem to indicate a less precipitous falling off in speed. Nonetheless, only the swiftest performances project even four beats per measure as opposed to eight, and a fair number project no beat at all. Flowing and sensuous, with interwoven vocal phrases and an all-woodwind backdrop, the adagio is an even better candidate for timeless beauty than the act 2 trio, "Ah taci ingiusto core." Of its twenty-two measures, only four contain figurations that strongly suggest a regular beat, a series of accompanying arpeggios in the clarinet occurring first as quadruplets and later as sextuplets. Otherwise the texture allows for almost limitless rubato and effacement of the pulse. The recordings in figure 7.2 render it at 80–90 bpm to the eighth, attenuating the beat with flexibility and lyricism but not obscuring it entirely. Others go the distance, reducing the eighth-note pulse to 60–70 bpm and rounding off every musical edge that might produce rhythmic definition. Examples include the recordings led by Fritz Busch, in 1936; by Hans Swarowsky, in 1950; and by Karl Böhm, Bernard Haitink, Rafael Kubelik, and Herbert von Karajan, among others, in the 1970s and 1980s. Having already set the maskers apart by decelerating at their D minor entrance, these conductors now separate them further. For one long moment, more than two and a half minutes in the slowest versions, the three conspirators seem to escape time itself.

And maybe the opera. If a "beautiful" act 2 trio interrupts a single instance of seduction, a "timeless" adagio pauses the plot itself, the whole busy spectacle of sex, violence, nasty tricks, and social intimidation. The maskers are praying that their hearts be protected or avenged, and their immediate worry is obviously the coming confrontation with Don Giovanni. But when the intrigue effectively goes on hold, the characters may seem equally concerned with the more existential challenge of being in Don Giovanni's story at all. Not coincidentally, perhaps, though the episode is not marked as an aside, it is often staged with the three characters addressing the audience directly. Looking beyond the garden, they appeal to the spectators for relief, or perhaps invite them to take their own respite from *Don Giovanni*'s cruel comedy.

Recordings at the other end of the rhythmic spectrum do not dispel all solemnity, but as I have written about another favorite for slow-motion transfiguration, the Count's plea for forgiveness at the end of *The Marriage of Figaro*, operatic prayers take on a different quality when they speak to immediate rather than existential circumstances.[27] In versions led by Norrington, Jacobs, Sigiswald Kuijken, Claudio Abbado, Daniel Harding, and Bertrand de Billy, all of which set the eighth notes of the adagio at 100–120 bpm, the prayer feels musically and dramatically integrated in a way that goes beyond any audible continuity of beat. Indeed, not all of them observe the usual 1:1 proportion between the eighth notes of the adagio and those of the preceding menuetto: in Kuijken's recording the relationship is more like 3:4, and in Norrington's and Harding's it is closer to 5:6. More important, their moderate and well-defined pulses keep the characters firmly in the plot, where indeed they have been all along. Compared to the examples shown in figure 7.2, which decelerate by 15–25 percent when the maskers first enter, those with the quickest adagios tend to keep the earlier transition steady (Norrington, Kuijken, and Jacobs), or to slow down by less than 10 percent (Abbado and Billy). They highlight the agitation of Mozart's D minor rather than its fatefulness or sorrow. The same disquiet invades the adagio, where Donna Anna's scales up to B♭, an occasion for soaring majesty at 70 or 80 bpm, are at 100 bpm more like fearful shudders, and where Donna Elvira's broken arpeggios, often ethereal at slower tempos, sound wandering and uncertain (compare Karajan 1985 [4:00–4:22] with Jacobs 2006 [2:36–3:01]). Even the arpeggiating clarinet may suggest rising unease when it shifts from quadruplets to sextuplets. The problem from which the characters seek deliverance, Don Giovanni, seems to remain very much on their minds, and while they may hope for succor, they cannot elude the comedy. Nor can they offer spectators much relief, at least to judge from live recordings: for many years an occasion for applause, the final cadence of the adagio is now equally likely to pass without fanfare. The otherworldly set piece has become an unsettling episode in its own right.

Everything Is Known

Between the prayer and the act 1 curtain lie seven more sections, the first two of which are themselves not unlike set pieces, especially the second, which stages a new, undeniably dignified entrance for the maskers, followed by shouts of "viva la libertà!" (table 7.4). Beginning without transition and ending with grand cadences, these are the most self-contained passages in the entire finale. By contrast, in musical as well as dramatic terms the remaining five sections all run together (indicated in the table by dashes). The menuetto breaks off at Zerlina's cry, passing without pause into the allegro assai, which in turn ends on an unresolved dominant chord leading onward into the andante maestoso. Another unresolved dominant concludes that section, and the ensuing allegro moves directly into the più stretto. The continuity brings matters of tempo planning to a head, as some conductors take the opportunity to engineer epic accelerations and others to enlarge on whatever seems to have transpired in the dance scene: hypnosis or struggle.

In its most elaborate form, accelerating to the end means taking progressively faster beats in four of the five concluding sections: menuetto (quarter

TABLE 7.4. Act 1 finale, second half

TEMPO MARKING	METER	CHARACTERS	EVENTS
Allegro	6/8	DG, L, Z, M	Party begins
Maestoso	2/4	Everyone	Maskers arrive, all sing "Viva la libertà!"
Menuetto –	3/4		Orchestras superimpose contradance and Teutsche; DG abducts Z
Allegro assai –	c		Z is rescued
Andante maestoso –	c		Guests confront DG
Allegro –	c		Confrontation and confusion
Più stretto	c		Confrontation and confusion

notes)—andante maestoso (quarter notes)—allegro (half notes)—più allegro (half notes).[28] In such scenarios, the very brief allegro assai becomes a kind of a parenthesis, its speedy half-note beats and agitated accents interrupting the longer-range progression of tempos only temporarily—for about thirty seconds, in most cases. Not surprisingly, the scale of the surrounding acceleration is largest among earlier recordings, where characteristically slow menuettos allow lots of room for speeding up. Figure 7.3 offers three examples from studio recordings, and there are similar live versions conducted by Walter, Klemperer, and Furtwängler. Even where the climb begins at a faster tempo, however, as in more recent studio recordings by Marriner, Barenboim, or Muti, the acceleration can be quite palpable. ♩

"The long, continuous line" of such performances, to recall Leinsdorf's phrase regarding his similar treatment of the *Figaro* finale, admits of at least two readings. On the one hand, the more dreamlike the first stage—the ballroom sequence—the more everything else sounds like sheer outrage. Jolted awake by the allegro assai, the libertine's antagonists subsequently go from indignation (andante maestoso) to anger (allegro) to fury (più stretto), their tempers rising with each increase in speed and musical intensity. As the later sections introduce effects such as chromatically ascending vocal lines, flying triplets in the violins, and crescendos paired with sudden pianos, Don Giovanni and Leporello seem to struggle just to keep up, wondering at the "horrible tempest" they have unleashed and eventually vanishing into the massive cadences of the più stretto. Almost a catharsis, the end of the act allows society to vent its rage.

On the other hand, the text also allows Don Giovanni a moment of defiance, sung just before the più stretto begins: "Ma non manca in me coraggio" (But courage is not lacking in me).[29] He is telling us that the story is not over, and that may be all that Da Ponte and Mozart intended, given that they did not add any stage directions. Directors have long turned words into deeds, however, calling on Don Giovanni to brandish a sword, scale a wall, or otherwise snatch victory from the jaws of defeat. With that in mind, it is tempting to hear strings of rising tempos as expressive not of society's rage but of the libertine's familiar energy—or at least, of his ability to coopt any rhythmic energies directed against him. The life force wins again, whirling out of reach and denying the others their best chance of exacting earthly justice.

He also sings his defiance in performances that do not accelerate so systematically, but here the effect differs, thanks especially to greater continuity between the menuetto and the andante maestoso—that is, between the

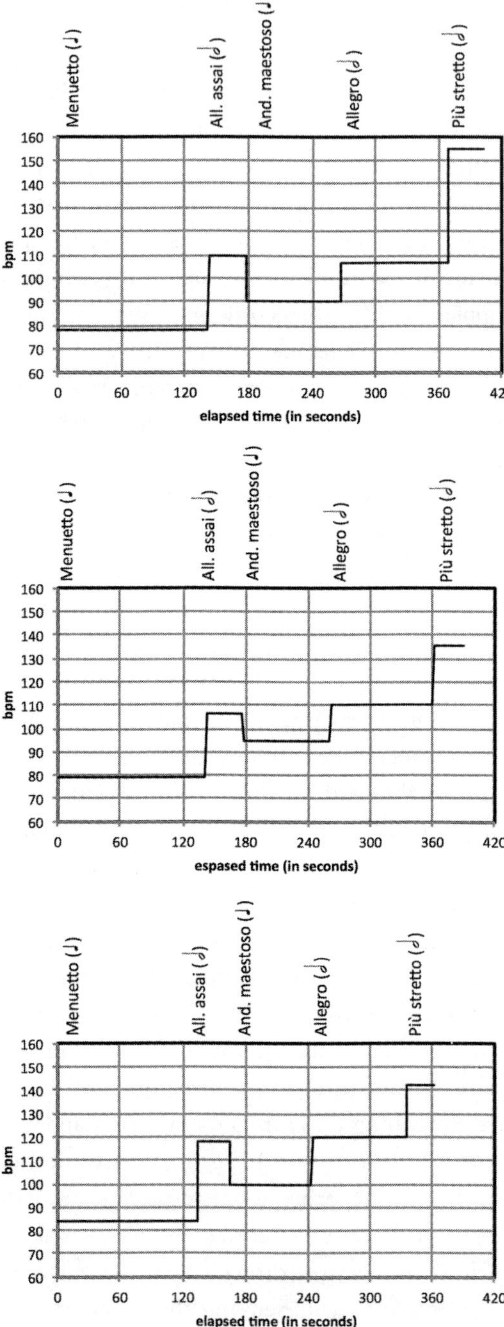

FIGURE 7.3 Act I finale: accelerating conclusions

a. *Don Giovanni*, cond. Busch 1936 (begins at 2:56)

b. *Don Giovanni*, cond. Krips 1955 (begins at 2:42)

c. *Don Giovanni*, cond. Rudolf 1955 (begins at 1:31)

ballroom sequence and the confrontation. As early as the 1950s, conductors such as Böhm, Fricsay, Giulini, Leinsdorf, and Hans Rosbaud were treating these two sections more like parallel tableaux than stages in an intensifying progression. They adopt virtually the same quarter-note pulse for both, shifting by such small margins that the associated actions seem to remain part of the same steady flow (fig. 7.4a). In later years an increasing number of conductors followed suit, keeping the pulse consistent until the allegro arrives, and in one case absorbing even the intervening allegro assai. As captured in 1999, Daniel Harding and his collaborators maintain roughly the same pulse of approximately 112 bpm for the entire sequence from menuetto through allegro, accelerating only at the più stretto (fig. 7.4b).

At slower speeds, continuity between the menuetto and the andante maestoso not infrequently suggests twin episodes of stupefaction. After the allegro assai disrupts the dance, the agitation lasts through the first business of the andante maestoso, Don Giovanni's effort to incriminate Leporello. Once the maskers reveal themselves, the music begins to settle down, and by the time Zerlina leads off a chorus of recriminations, "Tutto tutto già si sa" (Everything is known now, everything), a familiar frozen quality re-turns.[30] Over quietly pulsating strings—at 80–90 bpm, the upper strings' figure does not sound like tremolo—the accusers build a music-box texture of pointillistic eighth notes and mechanical dotted figures, all over the sim-plest of harmonies. Don Giovanni and Leporello are speechless for once, and though a modulation follows along with (traditionally) a crescendo, the others remain formal and restrained through the end of the section (Giulini 1959, 0:53–1:32). It is as if the tables were turned musically as well as dra-matically, and the protagonist, having manipulated time in the dance, finds himself victim to the same tactic. Accordingly, the very final sections, the allegro and più stretto, convey more surprise than catharsis: Don Giovanni finds his courage again and maybe even escapes, but the memory of his en-trapment lingers. Harnoncourt goes furthest in this direction, prolonging the sense of restraint even beyond the end of the accusations. Having per-formed the menuetto and andante maestoso at approximately 80 bpm, he sets the allegro at 105 bpm, still only as fast as some conductors' menuettos. He then accelerates mightily in the più stretto, all the way past 140 bpm, a leap so big that it sounds anything but inevitable (fig. 7.4c). Whatever we imagine to be happening, here or in other recordings that save their excite-ment for the end, it does not seem born of any long-building primal energy.

When the menuetto and andante maestoso share similar but faster speeds, the effect is different yet. The active, sometimes chaotic quality of

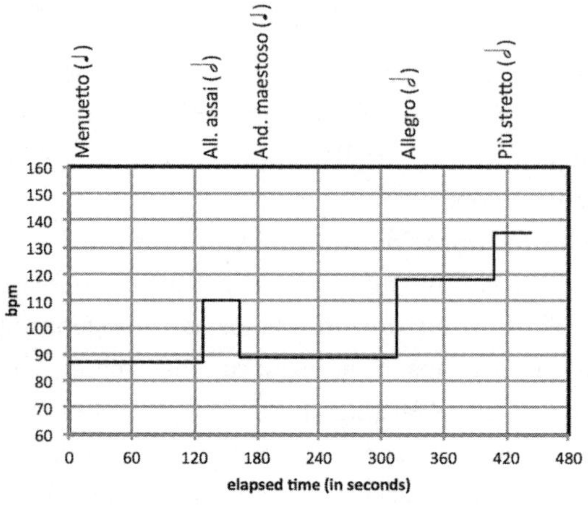

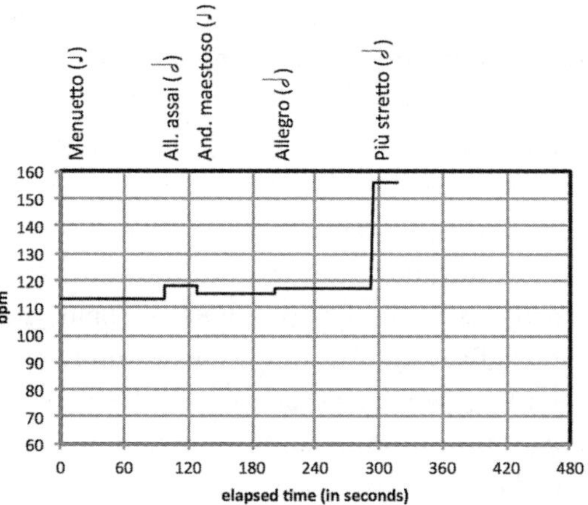

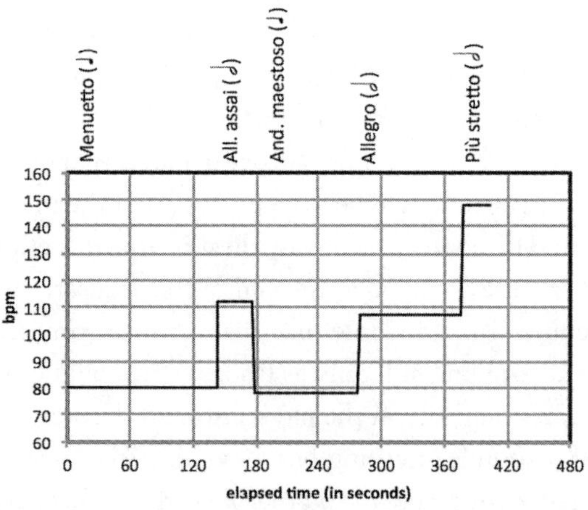

FIGURE 7.4 Act I finale: tableau-like conclusions

a. *Don Giovanni*, cond. Giulini 1959 (begins at 0:05)

b. *Don Giovanni*, cond. Harding 2000 (begins at 2:40)

c. *Don Giovanni*, cond. Harnoncourt 1988 (begins at 11:46)

the former carries over into the latter, finding expression in a sharp execution of the initial orchestral figures, in pointed renderings of all the dotted rhythms, and above all in a transformed delivery of the accusers' "Tutto tutto già si sa." At 95 bpm or higher, the upper strings' accompaniment becomes a tremolo after all, and the pointillistic vocal attacks are more suggestive of a game of darts than a music box (Harding 1999, 5:32–6:01). Having never succumbed to a spell in the first place, the characters do not spin one of their own but simply keep struggling against the protagonist, section after section. In Harding's recording especially, but also in Mackerras's, Gardiner's, Jacobs's, and others, the entire second half of the finale sounds like one giant imbroglio, a running battle that consumes everyone before, during, and after the defining act of violence. As time levels off, so does the playing field, and no character holds a supernatural advantage.

PART III
Partial Visions

Video Recordings

✳

CHAPTER EIGHT

Zooming In, Gazing Back

Few operas look as good on video as the 2007 *Don Giovanni* directed for the Netherlands Opera by Jossi Wieler and Sergio Morabito, and for Netherlands Television by Misjel Vermeiren. Barbara Ehnes's design divides the stage into modestly sized "bedrooms," mitigating the discrepancy between big stage and little screen (fig. 8.1). Decorated à la 1960s suburbia, the bedrooms suit a medium renowned for domestic drama, and so, too, does the story, retold as a clash between the sexual revolution (Don Giovanni) and the nuclear family (everyone else). This is opera as soap opera, so camera-ready that on video it is hard to tell where liveness ends and mediatization begins. A frequent demand of stage directors, critics, and opera fans seems finally to be satisfied: that cameras should document productions without interpreting them.[1]

Like the fidelities discussed in earlier chapters, though, *Treue* to the stage is an illusion. Filming and editing transform live opera, affording more but also more partial views than spectators enjoy in the house and guiding the attention via the choice and order of shots.[2] Video directors can make the action appear faster or slower: in the Netherlands example, numerous quick cuts animate a staging so static that one critic dubbed it a "concert performance."[3] Direction can also make singers appear more "in character" or "in performance" and the characters "deeper" or more "superficial." Here the Netherlands video meshes fully with its source, bringing the psychologizing devices of film to a stage production already rife with family dysfunction, repression, and other psychosocial dramas. If Mozart's music suggests emotions running beneath the surface, then close-ups, zooms, and

FIGURE 8.1 *Don Giovanni*, dir. Wieler and Morabito 2007: full stage with "bedrooms." DVD Kultur D0701, 2008.

other visual markers of interiority make his characters appear more human than ever.

The added insight comes at a price. Ironically, the Netherlands video is most faithful to the staging when it critiques visuality itself. Onstage, Wieler and Morabito have Don Giovanni or Leporello observing nearly every scene, the one glowering through black-rimmed spectacles, the other taping pornos with a vintage home-movie camera. Like allegories from the more Freudian reaches of film theory, they enact the male gaze before our eyes, asserting control over the others by turning them into visual objects. In the video, allegory becomes reality during the first encounter of Don Giovanni and Donna Elvira, when the image onscreen becomes shaky and discolored, as if we were watching through Leporello's handheld camera (DVD disc 1, 23:43; chap. 5, 3:27). The moment lasts only a few seconds, but it is long enough to remind us of the perils of spectatorship, especially as mediated by film. Even as cameras supply (some of) the visual dimension missing from audio recordings of opera, they implicate viewers in a potentially fraught exercise in operatic voyeurism.

The Styles of Opera Video

Part III focuses on opera video, my term for recordings that present complete operas in (usually) live theatrical performance. There are other kinds of filmed opera, dating all the way back to the earliest days of moving pictures. At the Paris Exposition of 1900, visitors could watch a clip of Victor

Maurel singing an excerpt from *Don Giovanni*, the music provided by synchronized cylinder,[4] and in subsequent decades the silent-movie opera became a cinema standby, accompanied by audio recordings or live performers. Sound films of opera followed, building to an apogee in the 1970s and 1980s, when many lavish examples appeared, including a full-length cinematic *Don Giovanni*, directed by Joseph Losey and released in 1979.[5] They preserve a rich vein of interpretation in their own right, creating wholly fictional worlds with studio or location shooting, prerecorded soundtracks, and postproduction editing. *Don Giovanni* has also inspired a number of cinematic adaptations, including retold versions of the opera itself as well as mock documentaries incorporating scenes of performance into fictional plots involving the performers.[6]

As souvenirs of performance, however, all of these are greatly outnumbered by the theater-based recordings that followed the advent of television in the 1940s, and especially of home video players in the 1980s. Their subject is as much the real-life work of enacting an opera onstage as it is the story of the opera per se, and the majority of examples from before 1990 originate in television series that broadcast uninterrupted season performances, such as *Live from the Met*.[7] Beginning with shots of the opera house and its audience, such presentations strove to give viewers the feeling of "being there," much like sportscasts but without the running commentary (though there is sometimes a host providing background and interviews).[8] They were rarely released as commercial videos, but private copies have long circulated widely, finding their way into classrooms and research libraries and more recently onto the Internet.

An alternative though related approach comes closer to cinema—or, better, to the studio-edited live releases of the audio recording industry (see chap. 5). Footage from multiple performances (live or mounted separately for cameras, with or without audience) is spliced together and edited into a continuous whole.[9] Originating in the 1950s and adopted widely after 1990, this technique undergirds numerous commercially available recordings, many of them produced as much for video release as for broadcasting. They nonetheless continue to imitate live television, showing the opera from beginning to end and including images of the performance event: house, audience, conductor, curtain calls. For the most part, directors and editors also avoid camera angles or special effects that would suggest anything other than a real performance unfolding in real time. After some initial experimentation, their aesthetic has carried over into opera's post-2000 return to movie theaters in series such as the Met's *Live in HD* and the Royal

Opera House's *Live Cinema Season*. Harking back to the original television broadcasts, they present and preserve a single performance on a single day, complete with hosts and onsite interviews.[10] The ethos of liveness endures.

Visual style is not so consistent, ranging from what I think of as a classic broadcasting approach, developed in the 1950s and still found today, to a more kinetic manner typical of post-1990 edited recordings. In the case of *Don Giovanni*, the difference can be seen right away in the introduction, a good test case inasmuch as it features several distinct situations in succession, like a miniature finale (table 8.1). Not unlike the conductors discussed in the last chapter, video directors shape the sequence according to both established convention and individual preference. The classic approach generally uses three or four cameras in standardized positions, such as rear center, or front right and left at stage level. Sections begin with wide establishing shots, followed by closer views marking vocal entries or stage actions—but not too close, the prevailing decorum calling for full or half bodies more often than heads or faces. The cuts between shots come at relatively wide intervals, generally every ten to twenty seconds, and cuts and shots are often shared across multiple filmings. Two versions show the conventions at work: the first *Live from the Met* broadcast of *Don Giovanni* (1978), directed by Herbert Graf and filmed by Kirk Browning, and an earlier version from the Komische Oper in Berlin (1966), directed by Walter Felsenstein and filmed by Georg Mielke (table 8.2). Both have cuts delineating the four main sections of the introduction,[11] wide establishing shots at the beginning of each section, and parallel editing within sections (e.g., the second section alternates shots of Don Giovanni and Donna Anna together, with shots of Leporello alone). Similar patterns occur in many other broadcasts, as do the twin approaches to the concluding andante, where

TABLE 8.1. Introduction

TEMPO MARKING	TEXT	METER	CHARACTERS	EVENTS
Molto allegro	"Notte e giorno faticar"	c	L	L complains of the servant's lot
	"Gente! servi!"	c	L, DG, DA	DG and DA struggle, witnessed by L
	"Lasciala, indegno"	c	L, DG, C	DG and C argue and duel, witnessed by L
Andante	"Ah, soccorso!"	¢	L, DG, C	C dies, DG and L react

TABLE 8.2. Camera shots in two broadcasts of the introduction

EVENTS	FELSENSTEIN/MIELKE 1966	GRAF/BROWNING 1978
	0:00	*0:00*
L complains of the servant's lot	4 shots: full stage, then alternate L full and upper body	4 shots: full stage, then alternate L full and upper body
	1:44	*1:35*
DG and DA struggle, witnessed by L	7 shots: DG/DA full body descending stair, then alternate DG/DA and L	6 shots: full stage, then alternate DG/DA and L
	3:13	*3:05*
DG and C argue and duel, witnessed by L	5 shots: zoom out to full stage, then alternate full stage, C, L	4 shots: full stage, then alternate C, DG, full stage
	4:17	*4:02*
C dies; DG and L react	6 shots: C, then alternate full stage, DG, L, C	2 shots: C, zoom out; DG, zoom out
Total	22 shots	16 shots

Browning pictures all three men together, and Mielke each one in sequence. The classic style favors regularly spaced shots keyed to the main turning points in the score and libretto.

More kinetic videos operate on a different set of values, employing more cameras and tending toward frequent cuts and tightly framed shots, including numerous close-ups. A second pair of recordings illustrates the difference, Vermeiren's from the Netherlands and a widely discussed video based on a Peter Sellars production of the late 1980s, filmed live in a studio in 1990 (table 8.3).[12] Again, the respective introductions have numerous similarities, including frequent cuts (the passage uses fifty-one to fifty-five shots as opposed to only sixteen to twenty-two in the broadcasts), heavy use of close-ups, delineation of numerous individual vocal entrances (e.g., the second section pictures Don Giovanni and Donna Anna separately, rather than together as in the broadcasts), and reaction shots (e.g., of Don Giovanni while Leporello sings).[13] The last device, so familiar from movies and television but virtually absent from classic-style opera video, has significant ramifications that will be explored below. More generally, the quick, intimate, and variable flow of images emphasizes the continuity of the scene over its demarcations. Where Browning and Mielke clarify the action, Vermeiren and Sellars embrace its chaos.

TABLE 8.3. Camera shots in two edited video recordings of the introduction

EVENTS	SELLARS/VOGEL 1990	WIELER AND MORABITO/ VERMEIREN 2006*
	0:00	*0:00*
L complains of the servant's lot	3 shots: L close-up (of back), then zoom out to L full or upper body	17 shots: alternate L and DG/DA
	1:47	*1:40*
DG and DA struggle, witnessed by L	20 shots: alternate DG/DA and L, also DG and DA separately	14 shots: alternate DG/DA and L, also DG and DA separately, C at end
	3:11	*3:09*
DG and C argue and duel, witnessed by L	14 shots: C entering, then alternate C, DG, L	12 shots: full stage, then alternate C, DG/C, DA, L
	4:13	*4:17*
C dies; DG and L react	18 shots: DG/C, then alternate DG, C, L	8 shots: C, then alternate DG, C, L
Total	55 shots	51 shots

* All characters onstage throughout

The two pairs of videos represent opposite poles of a continuum, and between them fall many distinctive approaches. The several *Don Giovanni*s of Brian Large, opera's most prolific video director to date, steer a kind of middle course, depicting the introduction in some forty-odd shots of moderate closeness. Claus Viller, filming a production by Michael Hampe at the Salzburg Festival (1987), uses many shots to capture singers dispersed across an enormous stage, but his perspectives remain distant. Despite a general trend toward quicker cuts and closer close-ups, moreover, practices in any given period may vary. In 1977, in an edited recording of the Peter Hall production for Glyndebourne, the director, Dave Heather, cut and framed in ways that anticipate Sellars and Vermeiren, while in much later versions from Madrid (2005) and Covent Garden (2008), Robin Lough adopts a classic style reminiscent of Browning and Mielke. Equally important, and notwithstanding the updated settings of both the Netherlands and Sellars productions (the latter takes place in 1980s New York), proliferating close-ups and cuts do not necessarily make a "director's video" to go along with "director's opera." One of the most active and close up of all *Don Giovanni* filmings, directed by Gary Halvorsen for the Met and using seventy-one shots for the introduction alone, portrays a period-dress staging by Franco Zeffirelli and Stephen Lawless, while Lough's comparatively relaxed pacing records a Lluis Pasqual production updating the story to 1940s Francoist Spain.[14] Whatever the prevailing aesthetic, onstage or on video, directors and their collaborators imprint each project with their own sensibilities.

Running Time

In *Don Giovanni,* nothing is affected more than the viewer's perception of time. Like its protagonist, the opera has something of a reputation for breathless energy, with the introduction as its catalyst. "An Introduction with such terse and violently moving action lies altogether outside the niceties of *buffa,*" writes Stefan Kunze, adding, "Even in other dramatic literature, parallels can scarcely be found in which the energy and motivation for the entire plot is drawn out of the first scene."[15] While the remainder of the opera includes the obligatory moments of reflection, the single-minded narrative—recall Allanbrook's "Don Juan is a libertine, so Heaven punishes him"[16]—unrolls with relatively few subplots or diversions. The director Jonathan Miller takes this quality as the very basis for interpretation: "If one does not resist the chaotic turbulence within which the characters swirl like human debris, and if . . . one submits to its spiralling momentum, the action and characterization take care of themselves. Everyone is swept towards the irresistible vortex of the opera's conclusion."[17] Some productions mount clocks onstage, and in the Netherlands Opera version Don Giovanni has an hourglass with which to time his seductions. Even with the action sublimated into psychodrama, everything still seems to course toward those final words of the statue, "Ah! there is no more time left to you."

Video direction can confirm the opera's urgency—or the reverse. The extended shot lengths of the classic style often act as brakes, imposing a dignified, pre-MTV tempo on both the visual and the musical events. Nearly always coinciding with the beginnings and ends of vocal entrances, the cuts in such filmings turn the interwoven textures of ensemble passages, in particular, into well-regulated exchanges, as in the back-and-forth views of the witness Leporello and the combatants Don Giovanni and the Commendatore in the introduction. Wide camera angles mitigate the intensity of the exchanges as well: the greater the distance, the less visible the frowns and sneers, and the less likely the physical struggles will range out of view. However violent their gestures, Don Giovanni, Donna Anna, and the Commendatore are always contained by the frame, their raging kept within bounds.

More kinetic videos may differ on some or all counts. Shorter shots increase the visual tempo, and cuts may establish their own rhythm independent of the vocal entries or musical phrases, complicating the temporal flow. Reaction shots do the same, imbuing the visual drama with some of the simultaneity of Mozart's ensemble writing. By showing every grimace and grunt, moreover, and by getting near enough that the characters sometimes flail offscreen, close-ups tend to maximize rather than contain the emotional energies. A flickering, multilayered, intimate sequence of images summons the "irresistible vortex" of Jonathan Miller's description.

Further influencing the perception of time is the relative emphasis on story versus performance. Opera video's investment in the latter is hard to exaggerate; even in examples that are taped separately from live performance and could theoretically avail themselves of all manner of cinematic technique, the model remains live television, with its multiple cameras pointed at a playing field or a studio set full of talking heads.[18] Still, visual style can influence both the literal and the figurative visibility of the live event. Nearer the classic end of the continuum, wider shot framings reveal the proscenium, the fourth wall, and sometimes the entrances and exits from the wings. Other conventions celebrate performance as such, such as the practice of framing aria soloists dead center, usually in half- or full-body shots, and shifting to full-stage perspectives for their final cadences. The event of the aria (especially the end, so often grand) becomes as important as its dramatic function, applying another kind of brake. *Don Giovanni* allows ample room for solo display, and the more the filming emphasizes the performative ebb and flow, the rhythms of entrance and delivery and applause, the more it may temper the dramatic momentum.

Nearer the other stylistic pole, quick cuts and tight framings nudge the attention toward the breathless world of the story. To put viewers "in" the action, at least in the way that cinema or television drama do, opera videos would need location shooting, traveling cameras, and the battery of devices used to encourage identification, such as point-of-view shots and eyeline matches. All of these figure in opera films, which are typically more immersive than even the most heavily edited performance-based videos. On occasion, though, the latter do get close enough that one might as well be onstage, and they can also make the images look more candid than composed. In an ensemble or recitative, seeing speaker after speaker at very close range can easily produce, if not immersion, then the illusion of being a bystander within the fictional space of the conversation. In *Don Giovanni* it helps that several scenes have an actual bystander, Leporello, and videos of the opera look most cinematic when he is observing the others. See him cowering or chuckling at one side, then see the protagonists through a camera positioned on that same side, and it is hard to avoid taking a subject position inside the drama. Together with shot tempo and the degree of motion onscreen, this tends to underscore the urgency of the story, edging viewers a little closer to the vortex.

Visualizing the Subject

Filming has still greater consequences for subjectivity, a more ambivalent domain in *Don Giovanni* than time. The differences between nineteenth- and early twentieth-century commentators over the nature of each character, which were discussed in earlier chapters, point toward larger disagreements about inner depth. E. T. A. Hoffmann described the protagonist as rent by internal conflict, and Hermann Abert saw unconscious motives lurking everywhere: Don Ottavio had his "manly affection," which absorbs him so thoroughly that he cannot rouse himself to action (see chap. 3), and even Leporello "is an entirely autonomous, independent figure of a wholly Mozartian kind, developing his inherent potential through his interaction with the other characters."[19] By contrast, Søren Kierkegaard and others saw Don Giovanni as a personification of eros or biological energies, and Victor Maurel regarded Zerlina as a representative of the "little people" who round out the libertine's litany of conquests.[20] The characters wear their identities on their sleeves and have no hidden potential to develop.

In more recent commentary, the Hoffmann-Abert tendency largely prevails; it is the rare writer, not to mention performer or audience member or opera blogger of the last hundred years, who does not treat Mozart's characters as real people. The most ambitious psychoanalyze them, beginning with Freud's amanuensis Otto Rank, who saw *Don Giovanni* in Vienna in 1921 and produced a book-length essay diagnosing the title character as suffering from an Oedipus complex. His final paragraphs draw a parallel with Mozart himself, arguing that the death of the composer's father only five months before the opera's premiere triggered "deeply ambivalent stirrings of affect" that gave him special insight into his hero.[21] Brigid Brophy and others elaborated the biographical connection,[22] and it entered pop culture through the 1984 biopic *Amadeus*, in which the deceased Leopold Mozart becomes associated in the composer's mind with the Stone Guest.[23]

Understood as a product of real-life repression, the opera seems destined to have characters with inner lives.

But not for everyone. According to Jonathan Miller, one reason to surrender to the opera's urgency is that the characters are ciphers: "Who *are* these people? Titles define the roles (Donna Anna, Commendatore, etc.) yet we never see the settings which would assign a social reality to them. Restless and unhoused, they project passions without displaying anything one would wish to call a *personality*."[24] Allanbrook, who elsewhere emphasizes Mozart's human insight, similarly distinguishes *Don Giovanni*'s characters as "statically conceived class-types rather than creatures of flesh and blood." Both writers draw a contrast with *The Marriage of Figaro*, whose intricate social and family dynamics generate a wealth of hidden agendas. Not so *Don Giovanni*, whose "radically homeless tribe" lacks backstory and a meaningful social structure.[25] The libertine is hardly Oedipus; lacking a reflective aria or even the consistent musical style that at least marks the others' social status, he is all surface and no depth, a blank spot in the middle of the cast. Allanbrook calls him "No-Man."[26]

On video, the inner life comes under the sway of two sometimes contradictory forces. One is television's reputation for psychological probity. As the media historian John Caughie writes:

> Since the early days of live studio drama, it had seemed that the television camera could approach inner realities in a particularly intimate way. The television close-up seemed to evoke the possibility of inner speech with an intensity denied to theatre and an intimacy and immediacy which the cinema could not achieve. The development of editing techniques and the fluidity of the filmed image seemed to promise ways of engaging the viewer in the instabilities and disorientations of dream states, fantasies and the surreal, and of making the unconscious material.[27]

Opera video has been described in similar terms. The video director Brian Large describes his own work as being "more intellectual, more psychological" than that of stage directors: "You can create more depth, construct more reactions, make characters available."[28] Writing about the Sellars videos, Marcia Citron notes the psychologizing overtones of close-ups: "With the typical shot from mid-chest through forehead, the videos court intimacy and draw the viewer into the innermost emotions of the characters."[29] One could add that even half- and full-body shots have a similar effect: in solos, making postures, gestures, or glances more easily visible can bespeak interior thoughts, while in ensembles, focusing on one character at a time marks each as an individual. To this extent, opera video shares in the intimacy and discernment attributed to television.

At the same time, close-ups of opera singers notoriously reveal their mechanics: breathing, shaping the mouth, moving the tongue, glancing at the conductor, sweating.[30] Cameras show the working professional as much as the psychic self. Which image wins out depends on how directors treat story and performance, a balancing act in which close-ups play a critical role—though not in themselves, necessarily, so much as in their context and content, and in the quality of their audio accompaniment. The last may well exert the greatest influence, especially over an opera's quieter moments. In a passage such as *Don Giovanni*'s introduction, with its dramatic contrasts and climaxes, even poor audio can arguably convey the gist: tempo changes, dialogue, high notes, the dissonant crash at the murder of the Commendatore. The effect is no worse, if also no better, than watching a television drama with a tinny orchestral soundtrack. In subtler moments, by contrast, inferior sound can be fatal, reducing the orchestra to an undifferentiated wash and obscuring vocal nuances. When its murmuring strings, delicate woodwinds, and intimate changes of tone are less than fully audible, a solo such as Donna Anna's act 2 aria, "Non mi dir, bell'idol mio," loses the very signs most suggestive of interiority.

A series of variables makes it difficult to map audio quality onto changes in filming style. While the audio signal used in television broadcasting and video recording has improved over the years, shifting from mono to stereo in the mid-1980s and from analogue to digital in the 2000s, the difference may be offset in playback. The low-fi speakers still installed in some televisions and computers can negate the digital and surround audio found on many newer videos, making them sound suspiciously like the old mono broadcasts. Conversely, by the 1970s many of those broadcasts offered terrific audio provided one listened not to the television but to a synchronized stereo signal transmitted over the radio, a modification not captured by the mainly homemade videotapes that have preserved such broadcasts for posterity. As in the era of short-playing records, audio quality depends significantly upon the equipment used by listeners, and the consequences for perception are no less profound. No matter how she looks onscreen, poor sound may make Donna Anna seem as unfeeling as her critics have made her out to be (see chap. 9), just as broad-spectrum richness may imbue her with emotional depth and complexity.

Divas and Dreamers

Video directors have more control over other elements impacting subjectivity, beginning with how and when they change shots. In quieter and more reflective scenes, some replace cuts with dissolves, one of the few special effects to be used with any regularity. By smoothing the transition from one image to the next, dissolves offer a visual analogy for the shaded musical contrasts of arias such as "Non mi dir, bell'idol mio" or Don Ottavio's "Dalla sua pace." Equally important, whereas cuts switch instantaneously from one perspective to the next, dissolves insert a momentary loss of focus, a blurring of the eyes, as if one were daydreaming. Strictly speaking, only the viewer experiences this effect, but in a context where the characters themselves look dreamy or pensive, it is easy to connect what one sees with what is happening in their minds. The flow of soft-edged images mimics the internal wandering from thought to thought.[31]

The impact of cuts depends on where and how quickly they occur. When timed to coincide with musical demarcations, such as vocal entrances or the beginnings and ends of phrases, they tend to make singers and characters appear self-possessed. Verbal, musical, and visual grammar all line up neatly, and any implied inner thoughts get channeled through phrase and section rhythms that are regular enough (at least in Mozart) to project mental coherence and discipline. By contrast, quicker and less predictable changes, especially those that interrupt musical statements, reduce the formality of the performance and the seeming assuredness of mental purpose. Contrasting videos of "Non mi dir, bell'idol mio" show the difference, which echoes in some ways the aria's own twofold identity as emotional outpouring and virtuosic showpiece. In the preceding accompanied recitative as well as the aria's opening larghetto section, Mozart applies his signature lyricism to Donna Anna's reproach of her fiancé, who has accused her of cruelty for postponing their marriage. A melting, diatonic melody flows out over murmuring inner strings, eventually joined by the *amoroso* woodwinds heard earlier in Don Ottavio's "Il mio tesoro intanto" (see chap. 3). Unlike that aria, however, "Non mi dir, bell'idol mio" also has an independent coda, marked allegretto moderato and pairing hopeful sentiments about the future with springing rhythms and bright vocal coloratura. Its pizzazz offended the Romantic sensibilities of Hector Berlioz, who wondered how Donna Anna, having delivered "a song of intense sadness, in which all the poetry of love finds vent in lamentation and tears," could seem "suddenly to have dried her tears and broken out into coarse buffoonery."[32]

Posterity has taken a kinder view, but still the contrasts pose a challenge, a potential conflict between expressivity and execution.

The 1978 *Live from the Met* broadcast offers one solution. The stage director, Herbert Graf, places Joan Sutherland's Donna Anna front and center, and her movements are limited to a few steps and turns, as well as a parting gesture in which she opens her arms wide in triumph. With similar economy, Kirk Browning uses a mere eight shots for the entire scene (a passage of nearly seven minutes), all framing the soloist in full or half body, and all timed to mark vocal or formal turning points: a high B♭ near the end of the recitative, the first vocal entrance in the aria proper, and the reprise of the larghetto melody (table 8.4). Nothing distracts the viewer from the musical design or from Sutherland's delivery, which Browning further celebrates with a heroicizing conclusion, a long zoom through the coloratura followed by a full stage shot for the arms-wide final cadence, exit, and applause. Here are, at once, a character sounding the emotional depths and a soloist commanding stage and house. Story and performance compete for attention.

The Sellars video resolves the tension differently. Against an American inner-city setting (see chap. 10), Dominique Labelle's Donna Anna exhibits clear signs of duress, pulling her hair, clutching at Carroll Freeman's Don Ottavio, rolling up her sleeve to reveal heroin tracks, and eventually shooting up. By the coloratura she is writhing on the floor, her inner torments made manifest for the audience to see. The video editing redoubles the effect (table. 8.4). The scene uses thirty shots, some of them punctuating the same moments as Browning's but most beginning mid-phrase or mid-measure, uncoupling the imagery from the musical structure: the coloratura, in particular, breaks into a flickering sequence with no rhythmic basis in the music. Thanks also to very tight close-ups and informal framing (partial faces, off-center positions, bodies moving out of view), the result is a hallucinatory stream of images matching Donna Anna's delirium. Rather than keeping her at arm's length, the cameras evoke "the instabilities and disorientations of dream states, fantasies and the surreal," to recall John Caughie's words. Berlioz would surely have hated it, but the video does its best to minimize the very inconsistency he found so troubling. The character looks no less feeling in showy music than in lyrical.

The Silent Treatment

Perceptions of interiority are further affected by images of silence. Once again this is especially true in quieter moments, where the visible mechanics

TABLE 8.4. Camera shots in two recordings of "Crudele!—Ah no, mio bene! . . . Non mi dir, bell'idol mio"

TEXT	GRAF/BROWNING 1978	SELLARS/VOGEL 1990
Recitative	*1:11*	*1:17*
"Crudele!"	3 shots: DA upper body, then DA/DO full body (m. 3) with zoom to DA upper body, then full stage reverse angle (m. 13, "abbastanza")	8 shots: DA/DO full body in aerial view (mm. 3–8), otherwise alternate DA, DO, DA/DO close-up
Larghetto	*3:24*	*2:54*
"Non mi dir"	1 shot: DA/DO three-quarter body	6 shots: alternate DA close-up and DA/DO upper body
	4:15	*4:10*
"Calma calma il tuo tormento"	Continuation of previous: DO exits frame, then zoom to DA upper body	4 shots: alternate aerial view, DA and DA/DO upper body
	5:08	*4:55*
"Non mi dir"	2 shots: full stage reverse angle, then DA three-quarter body with zoom to upper body	2 shots: DA aerial view, then upper body
Allegretto moderato	*6:09*	*5:54*
"Forse un giorno"	2 shots: full stage, zoom to DA upper body at vocal cadence (mm. 95–112); full stage for exit aerial view	10 shots: continue DA upper body, then alternate DA and close-up
Total	8 shots with 4 zooms	30 shots

of singing may appear more or less appropriate. In an extroverted outburst like Donna Anna's coloratura, the effort required for execution may seem perfectly in keeping with the mood, even when seen up close. In more reflective passages, the potential for conflict returns. The same character's high B♭ (on "Ab-ba-*stan*-za per te mi parla amore" [Love speaks to me sufficiently in your favor]), sung just before her aria begins, is at once poetic and technical, requiring an exposed leap into the highest register, usually performed *piano*.[33] Nearly every video director marks the moment with a cut or a close-up, which draws attention to the vocal challenge as much as to the emotional tenderness. It was presumably to avoid such tensions, and to endow the characters with more interiority in general, that the director

Jean-Pierre Ponnelle had the soloists stop singing in parts of his *Marriage of Figaro* (1976)—or rather, stop lip-synching, since they were filming in a studio to a prerecorded soundtrack.[34] In the Countess's "Porgi amor," another lyrical effusion, Kiri Te Kanawa wanders around a bedchamber full of sentimentally charged mementos, mouth closed. The results are two-fold: her voice appears to emerge directly from her mind, and her face takes on expressions that would be impossible to assume were she singing. The soliloquy turns into a dialogue, between an aural self within and a newly expressive visual subject without.

Though editing might allow it, performance-based videos are too committed to liveness to show singers reflecting silently when they are supposed to be singing. Nonetheless, directors do feature images of silence, picturing soloists when they are not singing, and other characters while they listen. Opportunities for the first option abound in Mozart's operas, where the orchestra frequently prefigures or repeats melodies sung by the vocalists. When a character is seen to reflect while the orchestra plays a melody, then to sing the same melody, it is easy to imagine a dialogue between interior and exterior, just as in Ponelle's "unsung" arias. Another "Non mi dir, bell'idol mio," staged by Peter Brook at the Aix-en-Provence Festival and filmed by Vincent Bataillon in 2002, offers a good example. As the orchestra plays the aria's opening theme, first in the recitative and then in the aria, the cameras focus on Alexandra Deshorties's Donna Anna. The second shot especially speaks with the clarity of a silent movie, as Deshorties turns to look at Mark Padmore's Don Ottavio, eyes glistening. When she actually sings the theme, it is as if something private has become public (fig. 8.2).

Images of nonsinging characters can work similarly, the interiority spreading from one character to another much as it does from orchestral to vocal passages. They occur frequently in *Don Giovanni*, where the characters are rarely alone onstage, leaving ample opportunity for reaction shots. In "Non mi dir, bell'idol mio," one often sees nearly as much of Don Ottavio as of Donna Anna, and his miming of sorrow, sympathy, or impatience signals an inner life no less rich than hers. Once the viewer sees him, moreover, it is natural to assume that she, too, is in the grips of feeling, even if she appears comparatively calm, or preoccupied with singing. The music strengthens the impression, performing the standard function of cinema underscoring by binding direct shots and reaction shots into a single time and mood. Whether the image is of Donna Anna singing or Don Ottavio brooding, both ride on the same murmuring stream of inner emotion.

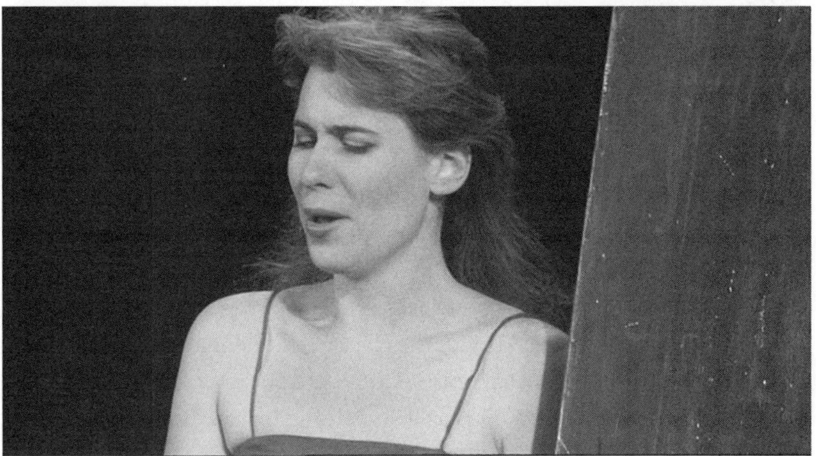

FIGURE 8.2 *Don Giovanni*, dir. Brook 2002. DVD BelAir Classiques BAC210, 2006.
a. Donna Anna (Alexandra Deshorties) looks at Don Ottavio as the orchestra plays the opening melody of "Non mi dir, bell'idol mio"
b. Donna Anna (Alexandra Deshorties) sings the opening melody of "Non mi dir, bell'idol mio"

Zooming In

Video's ultimate window on subjectivity, the zoom, features prominently in arias as well. Zooms suggest psychological insight by way of a physical sensation: as we draw closer and closer to a character, gradually rather than instantaneously, it feels as if we are journeying into the personality itself. Whatever barriers protect the inner life, the zoom drills through them, slowly and inexorably. A related device, the use of cuts to picture characters at successively closer range, occurs in a broader range of contexts, especially in classic-style videos, which treat all manner of scenes with establishing shots followed by more and more tightly framed views of individual figures. Less immediately suggestive than a zoom, the practice still implies a journey toward insight, even if the situation does not. As a revelation of the inner self, for example, Leporello's mini-aria at the beginning of the introduction bears little comparison to "Non mi dir, bell'idol mio": stomping around to a good-humored march and complaining about weather and food, the character seems all surface, a quintessential buffo. But as our view changes from the full stage to his body to his upper body or face, the last cut usually coinciding with his "Voglio far il gentiluomo" (I wish to play the gentleman), he looks increasingly like a real subject, with layers of personality.[35] The editing drives the statement inward, pushing it from comical aspiration toward deep-rooted desire. On video, everyone has an inner life.

This includes even Don Giovanni. When perspectives are wide and zooms and dissolves rare, he may indeed come across as inscrutable or lacking interiority; the baritone Mario Petri, for example, who tends toward icy self-confidence even in the presence of the Stone Guest, is not particularly humanized by the distant and static camerawork of the earliest video in which he appears, a broadcast of 1958. The production, Zeffirelli's first *Don Giovanni*, is notably warm and lively, but the filming reveals no cracks in the character's veneer.[36] Just two years later, though, in a recording taped at the studios of Radiotelevisione Italiana, the cameras work their magic.[37] Petri's characterization is no less steely, but close-ups shot in a controlled environment discover the little things that betray a racing mind beneath the surface—darting eyes, raised eyebrows, facial tics. Something similar happens to tough-guy Don Giovannis of later years, such as Samuel Ramey in the 1980s or Erwin Schrott in the 2010s.[38] If not exactly Oedipus, neither is this character "No-Man." Once they go looking, the cameras find as distinctive a subject in him as in anyone else.

Their success triggers another question, though: should they be looking in the first place? Borrowed from live television, opera video's affinity for the individual subject brings problems of its own. For one, highlighting the psychological runs the risk of obscuring other dimensions of the drama. *Don Giovanni*'s temporal urgency can be hard to detect in a video preoccupied with personalities, as can its social and moral overtones. Many videos ignore the overlay of three different dances in the act 1 finale, favoring the fragmented reactions of the individual characters over the composite tableau of different classes moving to different rhythms. This produces its own kind of tension, but in the realm of feeling more than of social relations. Equally if not more troubling, picturing characters as they reflect or emote—and in *Don Giovanni*, flirt and seduce and occasionally simulate sex—raises the issue of voyeurism noted at the beginning of the chapter. Plainly erotic imagery, the kind that occupies the porn-making Leporello of the Netherlands production, is only the tip of the iceberg. Opera videos have their share, but far more common are moments that evoke classic Hollywood cinema, whose subtler voyeurism is one of the linchpins of film theory. The male gaze, identified with the camera, captures women for its own pleasure, deriving an erotic charge from a feeling that the viewer is dictating the terms of encounter with a passive subject—or rather, object, since the woman trapped by the frame is deprived of agency.[39]

The theory has been modified to accommodate different genres and viewing habits, and opera video requires its own adaptations. Among other things, its cameras gaze at men as much as women, and at performers doing tangible labor as much as characters living out fantasies. Still, video does seem to objectify at every turn. It has helped to commodify singers' looks, enforcing frequently sexist standards of appearance that were once reserved for pop singers and film stars.[40] Video's love for the individual also leads it to fix on one person after the next, recording their every twitch and allowing them to be observed from afar, over and over. It is as if directors were trying to infect opera fans with scopophilia, the obsessive "love of looking" that psychoanalysis and psychoanalytic film theory regard as a pathology.[41] And with its male protagonist already objectifying everyone around him, *Don Giovanni* makes the gaze of the camera uncomfortably evident. One does not need a Freudian production like Wieler's and Morabito's to notice a parallel between the libertine's appraisal of the women onstage and our own appraisal of the singers onscreen.

Gazing Back

In the end, though, opera video may obstruct the gaze as much as enable it. Even in cinema, and even in films that construct the spectator as objectifying male, audiences can of course adopt alternative points of view.[42] In television the freedom is arguably greater, for viewers can watch not only from different perspectives but also with varying degrees of concentration, the television screen potentially constituting just one of several focal points in any given setting.[43] They may not watch at all, or tune in only when cued by the sound, an option undoubtedly exercised by more than one opera fan.[44]

Video allows yet more flexibility, rendering broadcast schedules irrelevant and permitting partial and repeated screenings. Powerful as they may be, the cameras never entirely determine what we see and hear.

Beyond that, the liveness so critical to television undermines what John Ellis calls "the voyeuristic mode":

> TV does not construct an image that is marked by present absence, its regime is one of co-presence of image and viewer. The image is therefore not an impossible one, defined by the separation of the viewer from it, but rather one that is familiar and intimate. . . . The cinema spectator is secure in the separation from the image that allows events to take place as though they were not watched. The broadcast TV image is quite often directly addressed to the viewer, in a simulation of everyday eye contact.[45]

Ellis's description accords with much of what happens in opera video. The format seeks not to distance its source but to bring it closer, to put viewers in the same room as the performers.[46] Rather than objects for private pleasure, the images are public, immediate, the common property of real and virtual spectators alike—or at least, that is how opera videos present themselves, with their shots of the opera house and other accoutrements of the live event. While singers rarely address the camera, moreover, they do sometimes engage the live audience, whether in asides or in arias played before the curtain. Such moments recall, if not the newscaster looking straight at us, then the talk-show host working the studio audience, a real person talking to real people in the same real world to which the viewer belongs. These images are not impossible ones.

The recording of opera's mental and physical feats works toward a similar end, lending the singers as much agency as the stars of live television, if not more. Beyond simple effort, cameras show what it takes to be an opera singer: magnificent sound, prodigious memory, talent for acting, capacity for split-second coordination. Indeed, opera soloists are such formidable subjects that rendering their abilities close up may well achieve something like the reverse of objectification. Carolyn Abbate argues that music "envoices" singers in opera, allowing them to claim precedence over the composer, to "shout out that *they* are creating the work literally before our ears (and eyes)."[47] Moving images of performers exert a related power, countering the dehumanizing constrictions of the camera frame with the human spectacle of action, professionalism, and engagement with the real world. If singers sometimes languish under the gaze, they also gaze back.

And they do so even in the most unpromising situations. An aria like "Non mi dir, bell'idol mio" seems tailor-made for voyeurism, with its female soloist lost in vulnerable reflection. Yet it can elude the gaze as well, thanks to the force of the performance and to a final characteristic of filming, a tendency to betray its own, technologically enabled agency. The very things that make viewers feel in charge may also reveal their lack of control: multiple cameras afford multiple perspectives but only as the editor sees fit; close-ups foster intimacy but shrink the view and force it in particular directions. Zooms are still more ambiguous, objectifying like no other kind of shot but also undermining the identification of the camera's eye with our own. Unlike a tracking shot, which moves as if the camera were a character, the zoom acts like a set of binoculars, sometimes to bizarre effect. In more than one "Non mi dir, bell'idol mio," Don Ottavio slides bit by bit off the screen as the camera homes in on Donna Anna, sometimes disappearing altogether and sometimes remaining as a vestigial, greatly enlarged ear or hairline. Small wonder that some film scholars characterize the zoom as having a distancing effect, offering an intimacy that is all too clearly artificial.[48]

Which returns us one last time to the media-savvy *Don Giovanni* of the Netherlands Opera. Other productions show better how singers and film technology may undermine the camera's gaze. In the Met's 1978 "Non mi dir, bell'idol mio," Joan Sutherland evinces far too much prowess, and the editing too much agency, for there to be any illusion that she is "passive" and we are "active." On the contrary, after four big zooms in only seven shots, it may be viewers who feel manipulated, like riders strapped into a visual roller coaster. Filmed in a less heroic style, the Netherlands "Non mi dir, bell'idol mio" looks more like the old Hollywood films that inspired the original theory of the gaze, capturing Myrtò Papatanasiu's Donna Anna alone in her bedroom writing a letter to Don Ottavio (fig. 8.3). Shots of other characters show that she is indeed the object of one man's thoughts and another's sight: Don Ottavio (Marcel Reijans) writes his own letter to her, and Don Giovanni (Pietro Spagnoli) looks right at her through his menacing spectacles. Midway through the virtuosic coda he even approaches, and a camera positioned close to the stage approximates his point of view (DVD disc 2, 1:06:27; chap. 12, 7:07). More than just depicting a woman as a sexualized image, the video invites us to identify with a character doing that very thing onstage.

FIGURE 8.3 *Don Giovanni*, dir. Wieler and Morabito 2007. DVD Kultur D0701, 2008.
a. The camera gazes at Donna Anna (Myrtò Papatanasiu)
b. Donna Anna (Myrtò Papatanasiu) gazes back

Yet the moment also resists objectification. While less vertiginous than in the Met video, a series of zooms again compromises whatever illusion we might have of seeing Donna Anna revealed. Our insight is too obviously controlled by the lens's mechanism, and by the editorial agency that aligns each zoom with a formal division: beginning, second verse, reprise. Just when the camera has us look through Don Giovanni's eyes, moreover, two further events dispel whatever trace remains of Donna Anna's vulnerability—or of Papatanasiu's. First, she launches into the final cadences, whose broken phrases and spectacular leaps demonstrate her vocal prowess even better than the preceding coloratura. Second, she literally gazes back at

Don Giovanni, her eyes meeting his in a gesture combining supplication and defiance (fig. 8.3). The gambit fails, Don Giovanni remaining still and Donna Anna sinking back to her bed. Yet no sooner does she do that than the soundtrack dissolves into applause, confirming the agency of the singer just as the challenge to Don Giovanni's gaze—and by extension, to ours—does that of the character.

CHAPTER NINE

Trauma Retold

When Donna Anna enters in the introduction, she rushes onstage, calls Don Giovanni a scoundrel, and vows never to let him escape. Later, she tells Don Ottavio why: after repulsing an assault in her bedroom, she follows the libertine outside to stop him, but he escapes after all, murdering her father and plunging her into an emotional tailspin from which she never recovers—not by the standards of opera buffa, anyway. Though she confirms her love for Don Ottavio several times, in the end she asks to postpone their wedding while she recovers. If she ever gets her happy ending, we do not see it.

In performance and criticism alike, her hesitation, her abiding sorrow, and her fury have long sparked controversy. Some see the realistic, even necessary reactions of a woman reeling from trauma; but others suspect a hidden motive, a conscious or unconscious desire to be Don Giovanni's lover, free of both father and fiancé. In criticism, the second view originated in the nineteenth century and had its heyday in the mid-twentieth, contested all along and eventually pilloried as sexist. In performance, the chronology is effectively reversed, at least on video, where a "conflicted" Donna Anna appears first in the late 1970s and with increasing frequency thereafter. Her emergence points up two paradoxes. On the one hand, the more degenerate Don Giovanni of the 1970s and after does not always make for more righteous antagonists, as we shall see in the next chapter as well. Sometimes the rot just spreads, and everyone languishes in moral ambiguity. On the other hand, the urge to uncover repressed feelings, a comparatively modern phenomenon in Mozart staging, can result in some surprisingly premodern characterizations. With her hunger to indulge forbidden passions, the conflicted Donna Anna has a lot in common with a woman E. T. A. Hoffmann claimed to know.

Analyzed

It was Hoffmann who invented Donna Anna's desire for Don Giovanni, in a passage that fairly revels in her supposed fall: "She was not saved! When he fled, the deed was done. The fire of a superhuman sensuality, an ardor from hell, flowed through her heart and made all resistance futile. Only *he*, only Don Juan could kindle in her the lusty insanity with which she embraced him, he who sinned with the overpowering, destructive rage of the infernal spirits within him."[1] Hoffmann admits that Donna Anna's lust changes to hatred, but in his view, this is because she is spurned, not traumatized. Several corollaries follow, including that the "frigid, vulgar, faint-hearted" Don Ottavio is no match for her ardor; that the same ardor will kill her in the end; and that the other women pale by comparison. The last two ideas had consequences onstage, in the Parisian adaptation of 1834 that ended in her funeral (see the introduction), and in the performances of the nineteenth-century soprano Wilhelmine Schröder-Devrient, who is sometimes credited with having made Donna Anna into the opera's prima donna.[2] To judge from a contemporary account, she read her Hoffmann carefully: "As Donna Anna she was fully the woman who succumbed sensually to the power of the man, and whose pain was only the cry of an unsatisfied infatuation . . . [audience members] could not help but believe that her passion for Don Juan, perished in the flames, made her recoil from her union with her conventional, unimpeachable, boring fiancé."[3]

Hoffmann does not limit himself to character analysis. As noted in chapter 3, he couches his story as an account of a live performance, and where Donna Anna is concerned he repeatedly muddles fact and fiction—or rather, "factual" narrative and "fictional" opera. At her entrance, character and singer alike come under his gaze:

> She could have been taller and slimmer, and more majestic in her movements: but what a head! Eyes from which love, anger, hatred, and despair fly out like sparkling beams from the *single* focal point of a prism, and which like Greek fire set the heart inextinguishably alight! The loosened braids of her dark hair fall in curly waves down the nape of her neck. Her white nightgown treacherously reveals charms that are never observed

without temptation. Seized by the horrifying deed [Don Giovanni's assault], my heart pounds with powerful blows. (17)

In due course she visits Hoffmann's box, praising his opera *Undine* and
triggering in him a recurring fantasy of being caressed. During her final solo, the rhapsodic "Non mi dir, bell'idol mio," he says, "I felt myself
shuddering in drunken rapture from a soft, warm breath that glided over
me. My eyes closed involuntarily, and a glowing kiss seemed to burn on my
lips—only the kiss was like a long-held tone of eternally thirsting yearning" (22). Similar sensations arise twice more after the opera, including
at the hour when life imitates art and the singer dies (24, 28–29). Donna
Anna, it seems, is not the only one with mixed motives; Hoffmann himself
means not merely to interpret the character, but to become part of her story.
His narrator seeks to replace Don Giovanni as the object of her desire.

Finding no supporting evidence in the opera's texts and troubled by his
impugning of feminine virtue, nineteenth- and early twentieth-century
writers such as Alexandre Oulibicheff, Otto Jahn, and Hermann Abert
parted ways with Hoffmann on Donna Anna, much as they had on the
protagonist (see chap. 3). In Abert's words, "Her loathing of a man who,
in murdering her father, dealt her, too, a mortal blow, consumes all her
other feelings. This hatred is not, therefore, a concealed form of love, as it is
with Donna Elvira. It is not a secret and self-tormenting desire to recapture
love's bygone pleasures but a desire for the blood of the criminal and, as
such, a wish to continue the game of life and death that he himself began."[4]
An "austere and noble figure," Donna Anna is reticent with Don Ottavio
only because of her aristocratic self-control, and the performance tradition
inaugurated by Schröder-Devrient is mistaken: "Still less does she react in
the manner of a vengeful Armida, so that it is wrong to cast a *hochdramatisch* soprano in the part, as Mozart intended her as a young woman whose
strength of purpose is tempered by maidenly tenderness and shyness."[5]
Like Zerlina, Donna Anna is in the end respectable, free of sexual desire
and any unfeminine thirst for power.

Hoffmann found a more sympathetic ear in psychoanalytic commentators, beginning with Otto Rank, who subsumes Donna Anna into the
oedipal drama he sees playing out between Don Giovanni and the Commendatore (see chap. 8). Dominated by her father but also connected to him
by "strong libidinal ties," she "partly welcomes the murderer as a liberator
and a new beloved, and partly scorns and persecutes him as a weaker substitute for the lost primal object."[6] Their relationship need not be consum-

mated to be psychologically motivating; real or imagined, Donna Anna's divided loyalties determine her state of mind and her actions. The novelist and poet Pierre Jean Jouve picked up the thread in 1942. Unlike Hoffmann, he respects the character's account to the extent of calling her unseen bedroom encounter with Don Giovanni a rape. Nonetheless, he also declares that she is at fault, a familiar accusation in rape cases that here takes the form of a diagnosis. "This woman, unconsciously attached to her father, has transferred to the man [Don Juan] her desire to be seduced, to be forced, at the same time that she feels for the seducer, unworthy rival of her father, a love-hate of which she can never be healed. She summons the rape, and the rape causes the death of the one she loves, and she in turn dies."[7] Her quest for revenge is driven by guilt, and her coolness toward Don Ottavio by a self-destructive hate that claims her will to live. Worse still, she suffers alone. Now sounding more like Hoffmann, Jouve claims that "Fin ch'han dal vino" reveals "that which we are fated to be: a diabolical unconscious . . . Don Juan owns and loves the whole room. There is a hysterical communion that no mind can deny—under the sign of repetition, we are possessed by expansiveness, by mythical wonder."[8] As an ersatz father, Don Giovanni vanquishes Donna Anna, but as a pseudo-Mephistopheles, he frees everyone from the mundane.

Jouve would eventually be rebuked on his own psychoanalytic terms. In the meantime, a variety of more ad hoc interpretations circulated: that Donna Anna was seduced and forced to lie about it;[9] that she was neither seduced nor raped but awakened to conflicted feelings;[10] that she was engaged in a fully consensual, "passionate love-affair with the Don."[11] The author of the last claim, R. B. Moberly, supports it with an enthusiasm for sexual conquest unusual even among the libertine's admirers, crediting him with as many as seven successful seductions in the time of the opera alone, libretto notwithstanding. No longer dressed up as Romantic fantasy or psychoanalysis, his skepticism of Donna Anna veers toward misogyny: "Briefly, she is an impulsive wilful motherless rich girl."[12] Harsher yet is a text from 1977 by William Mann, which accuses her of having "etiquette where her feelings and brain should reside."[13] Thus far, Mann recalls Edward Dent, who complained in his *Mozart's Operas* of 1913, "I suspect that if [Donna Anna] had not been brought up from childhood always to conceal her real feelings, and never to acknowledge to herself any motive but duty and family pride, she might have been much the same sort of person as Fiordiligi in 'Così fan tutte,' and there seems every probability that she will eventually become first lady-in-waiting to the Queen of Night."[14] For anyone who missed the point, Dent's second edition of 1947 adds: "She is in fact a thoroughly unpleasant young lady."[15] Mann outdoes even these expressions of disdain, inverting the Freudian narrative to arrive at another ugly stereotype: "All men, to her, are beasts, and it would be beneficial to her personal growing-up if she had been pleasantly raped by Don Juan (as some writers on Mozart's opera assume, I think optimistically)."[16] Diagnosed by Jouve of desiring rape unconsciously, Donna Anna is now taken to task for not desiring it enough.

None of these views was universal: record booklet commentators of the 1950s through the 1970s, for example, rarely portray Donna Anna as either puritanical or volcanic.[17] By the late 1970s, however, doubts as to her truthfulness and capacity for feeling enjoyed enough currency that writers such as Charles Osborne and Julian Rushton felt obliged to defend her.[18] More consequentially, Catherine Clément went on offense, rehabilitating all the women in the opera by cutting their would-be lover down to size. Amid rising calls to recognize the pervasiveness of sexual assault—Susan Brownmiller's *Against Our Will* appeared in 1975—Clément reminded readers that Don Giovanni relies more on force than on persuasion: "Twice the 'bird in hand' [Zerlina] goes along with the intoxicating words and subtle music. When Don Giovanni finally manages to drag her into a separate room and starts doing things, there is a scream. The same scream that Donna Anna cried out in the darkness. . . . Yes, at the end of the charming seduction is rape, its underlying violence."[19] Taking the libretto at its word, Clément further notes that even force does not spell success: Donna Anna and Zerlina both fight Don Giovanni off. *Contra* Jouve, then, to Clément's mind the real hysteric is the libertine himself, who brags about a promiscuity that exists only in his mind, and whose sexual adventurism brings only misery. To say the women are drawn to him amounts to little more than victim-blaming: "Maybe Donna Anna was never the object of an attempted rape; no doubt Don Giovanni went to her room to talk politics, because it seems he represents the revolution also. Maybe Zerlina cries over nothing. No doubt they are victims of their fantasies. Maybe all the evidence always given by cops about the obvious desires of raped women is true too."[20] In Clément's telling, *Don Giovanni*—or rather, its myth—does not liberate anyone. On the contrary, it leaves women at the mercy of sexual aggression and its apologists. What Hoffmann began by imagining a kiss, other commentators carry on with their stories of women filled with lust.

Clément's counterattack prefigured the reevaluation of the opera discussed in earlier chapters. Not everyone was convinced: in 1987, Julia Kristeva argued that "one need only have Mozart's joyful and stately music ring out above that edifying story, born of a medieval morality play in a state of decay, and the entire point of view changes; instead of the sullen claims of the victim, the air resounds with the pure jouissance of a conqueror."[21] Within a few years, though, Stefan Kunze and Paul Griffiths would treat the "conqueror's" offenses as exceeding even Mozart's ability to redeem them, and Charles Ford would describe both opera and myth as symptoms of a misogyny originating in the composer's own time: "Within the Enlightenment's conception of femininity, the complicity of every rape victim must always be in doubt, for no woman can *naturally* resist the sexual intentions of a man, for she is nothing more than the object of his desire."[22] The turn of the twenty-first century brought fresh scrutiny of claims about the opera's women. Noting that previous iterations of the Don Juan legend do in fact represent her as attracted to the libertine, Kristi Brown-Montesano argues that Mozart and Da Ponte intentionally take the opposite tack, giving no evidence of illicit desire and plenty of genuine feeling: "From her first entrance she staggers and bolts into emotional extremes that are fully justified by the tragic events of her life," and her commitment to Don Ottavio is amply evident in "Non mi dir, bell'idol mio."[23] In a related vein, Liane Curtis criticizes textbook accounts of the 1980s and 1990s, arguing that they promoted their own "strong Romantic hero" by emphasizing Don Giovanni's bravery and charisma, and by excerpting only those numbers that showcase his ostensible skills with women, notably Leporello's "Madamina, il catalogo è questo" and the duettino for Don Giovanni and Zerlina. Ignoring the rest of the opera conceals its violence, and with it the bravery of Donna Anna, whom Curtis judges "a towering figure of female strength . . . [who] has both fought off Giovanni's physical attack and struggled to keep him from escaping."[24]

The character's journey is not over. Critical and scholarly accounts no longer have Donna Anna desiring Don Giovanni's assault, but textbooks continue to favor the Catalogue Aria and the duettino even as many have revised their descriptions of Don Giovanni.[25] Apparently, neither the steadfastness of "Non mi dir, bell'idol mio" nor the rage of "Or sai chi l'onore" satisfies expectations of Mozartian realism, at least as it concerns women. Zerlina and Donna Elvira are deemed human for saying both "yes" and "no," but Donna Anna, the one who says "no" and sticks to it, has yet to gain everyone's trust.

Dignified

Turning to performance, the first thing to say is that many of the productions captured on video leave no doubt that Donna Anna tells the truth. They portray all of her reactions to Don Giovanni as sincere, including her fury during their initial struggle, her ignorance of his identity when they meet again later in act 1, and her shock when she recognizes his voice. When she then addresses Don Ottavio, both her narrative and her call for vengeance—the recitative "Don Ottavio, son morta!" and the aria "Or sai chi l'onore"—confirm her veracity. Characterizations in this vein are found from the 1950s through the first two decades of the 2000s, crossing epochal changes in musical and staging practice.

In the early decades of video, Donna Anna's conviction is buttressed by the decorum of mid-century opera staging, with its typically grand costumes and statuesque poses.[26] In two versions of the narration scene, both filmed in 1960, Teresa Stich-Randall stands firm throughout, delivering recitative and aria alike with an almost steely composure. Only at the end does she go into motion, sweeping grandly offstage and leaving Luigi Alva (Don Ottavio in both cases) in respectful wonder.[27] More typically, productions mix dignified stasis with signs of vulnerability. In a 1961 video from the Deutsche Oper, Georges Wakhevitch's stage design features a towering seminude female figure, an icon of Don Giovanni's dissolution, looking down over a distorted vista of columns and arches—a neoclassical villa gone awry (see fig. 9.1d).[28] Underneath her on a stair, Elisabeth Grümmer's Donna Anna stands in protest. Her gown (also by Wakhevitch) combines dark drapery with a white whisk collar, concealing the body and accenting the face; and Carl Ebert's direction holds her erect and largely stationary. Physically composed and guarded by God (she wears a cross), she projects all the order and virtue that the surroundings lack.

The vulnerability of this upright figure emerges in small gestures and facial expressions, as well as in musical effects. Donna Anna's narrative begins on a mysterious E♭ minor chord in the strings, marked by Ferenc Fricsay (as by other conductors) with a ritard and pause, and by the camera director, Rolf Unkel, with a shift from wide-angle to half-body view (fig. 9.1a; DVD disc 1, 53:10; chap. 22, 1:03). The stage is set for a recollection, and Grümmer sets out looking and sounding reflective. Within a few phrases, though, she veers toward reenactment, presumably encouraged by quickening rhythms and increased orchestral activity, and by shifts between past and present tense in the text (both the original Italian and Georg

FIGURE 9.1 Donna Anna (Elisabeth Grümmer) and Don Ottavio (Donald Grobe), "Don Ottavio, son morta! . . . Or sai chi l'onore." *Don Giovanni*, dir. Ebert 1961. DVD Arthaus Musik 101574, 2011.

a. The narrative begins

b. Donna Anna escapes

c. "Or sai chi l'onore"

d. Exit and applause

Schünemann's German). Even during the E♭ minor chord, her right hand reaches forward as if grasping for help, and as the text goes on to describe a man entering Donna Anna's room, she reaches further while also growing louder and straightening her posture: the character summons courage amid her terror. Each subsequent event prompts similar representations, with Grümmer's voice ranging from cries to whispers and her gestures from clenched fists to side-to-side twisting, the last at the moment Donna Anna describes her escape: "daß ich mich durch heftiges Wehren, Ringen, und Entwinden, endlich befreite" (that through violent fighting, wrestling, and twisting away, I finally freed myself; fig. 9.1b). Upright determination notwithstanding, her past threatens to disrupt the present.

Full of rage and marching rhythms, "Or sai chi l'onore" ("Du kennst nun den Frevler," Now you know the wrongdoer) signals restored confidence, and Grümmer sings with all the sonorous power, rhythmic steadiness, and unornamented clarity of mid-century modernism. A complementary return to wide-angle filming shows both character and singer commanding a broad expanse of the stage, her outstretched hand now conveying imperiousness (fig. 9.1c). But Mozart's music incorporates his signature contrasts, notably in the second verse, where a memory of the Commendatore's wound and blood elicits lyrical phrasing and a departure from the bright tonic of D major, a passage that is echoed just before the heroic final cadences (disc 1, 57:14; chapter 22, 5:07). In both cases Grümmer harks back to the terrified whispering of the recitative, and in the second her posture crumbles as well, leading to the first moment of contact with Donald Grobe's attentive Don Ottavio. This prepares the very end of the number, where the orchestra follows the vocal cadence with one of its own, ending in sustained chords that are unlike anything else in the aria (disc 1, 58:07; chap. 22, 6:02). Often the occasion for some further detail of staging, in this instance they accompany more evidence of weariness, as Grümmer's character covers her face and slumps backward; then is caught by Grobe's, who ushers her off the stage (fig. 9.1d). Dignified to the end, and in one sense triumphant—the exit is accompanied by applause—the performance nevertheless registers the double toll of experiencing violence and reliving it.[29]

Devastated (and Alone)

In later decades, the twin advents of a grittier Mozart (historically informed or otherwise) and more naturalistic stage conventions raise the emotional pitch of Donna Anna's testimonial. Singers always tended toward astringent timbres and sharp accentuation, but new details highlight the contest between vengeance and remorse, especially in the aria where "hardening" appoggiaturas ("Or sai, chi l'o-*no*-re") distance the first verse still further from the second. Changes in stage practice reach further, especially the post-1980 turn toward habits of acting modeled on spoken theater and cinema.[30] By asking singers to animate more of their bodies than had been typical in the 1950s and 1960s, as well as to move more fluidly, directors opened different realms of expression and arguably tilted the emphasis from singer toward character, performance toward story. Decorum had served in part to facilitate good vocal production, but dynamism tends to privilege action and emotion, and while sopranos still draw applause at the end of "Or sai chi l'onore," their enactments appear to follow dramatic cues as much as performative conventions.

Acting changes independently of the more controversial innovation of the 1970s and 1980s: updated costumes and settings. Even the most traditionally attired Donna Anna may sit, kneel, or otherwise abandon upright composure, and she may also use physically or sexually graphic gestures within her narrative. Edita Gruberová's performance at La Scala in 1987 takes a step in this direction: while luxurious costumes and a neoclassical set (not awry this time) locate the action in the eighteenth century, Giorgio Strehler's direction has her rushing down a staircase, crossing the stage, crossing back, and crumpling to the floor, transforming the recitative into an anxiety of motion. Decorum reigns only in the aria, delivered with renewed authority from a classic pose.[31] A decade later in Ferrara, in a production set in the nineteenth century and directed by Lorenzo Mariani, Carmela Remigio's performance is still more active. She reenacts the whole assault on the floor, twisting and crawling and eventually collapsing at full length to illustrate its violence, and during the aria she strides back and forth, sits, stands, and embraces Don Ottavio. Her one traditional posture—feet planted, fist raised—looks almost retro, as if she were memorializing a dignity that the modern Donna Anna can no longer maintain.[32]

Relocating the opera to the twentieth or twenty-first century does not enhance naturalism so much as localize its associations. As Mary Hunter points out, opera has always incorporated "familiarizing" touches into its historical settings.[33] The 1961 Deutsche Oper video discussed above offers a good example: Grümmer and Grobe sport mid-twentieth-century hairstyles to go with their Renaissance-style finery, reassuring audience members that their characters are not so foreign after all. At the same time, such connections quickly become dated, and to modern viewers, Grobe's bouffant and sideburns look just as historical in their own way as his doublet and sword. Updating an entire production is similarly double-edged. As Hunter writes of the modernized *Don Giovanni*s directed by Peter Sellars and Calixto Bieito, "These productions rely on the familiarity—at least in principle—of their settings. In other words, even though many audience members will not actually have spent time in the 1980s South Bronx or the bars of Porto Olimpico [in Barcelona], they 'know' from the media what life is like there, and they can instinctively supply a whole set of social particulars that flesh out the opera's action."[34] Those same particulars imprint naturalist movement with markers of specific times and places. When she falls to the floor during Sellars's version of "Non mi dir, bell'idol mio," Dominique Labelle recalls the performances of Gruberová and Remigio, except that the reinterpretation of the floor as a New York City sidewalk, and of Donna Anna as a heroin addict, turns the collapse into a metaphor for American social crisis around 1990 (see chap. 8). Modernization distances the action even as it brings (or brought) it closer.

Among naturalist Donna Annas who are also truthful, whether modernized or not, the common thread is a deepening vulnerability. Unable to retreat into dignity, the women played by Labelle and Remigio appear all but overwhelmed by their multiple traumas. Still more distraught is the Donna Anna played by Hillevi Martinpelto in a myth-busting production directed by Deborah Warner for the 1995 Glyndebourne Festival.[35] Dressed as ordinary professional and middle-class types, the characters range around an abstract set designed by Hildegard Bechtler, a stark, open space that hosts vivid and disturbing action: Donna Anna and Don Giovanni struggle fiercely at the beginning, and near the end he assaults a statue of the Virgin Mary in a mock rape. As Jean Andrews writes, "Warner finds precious little glamour in [Don Giovanni's] psychological profile . . . especially as it manifests itself most often as violence directed against women, a joke or a compulsion she finds no longer acceptable. Hers is not so much a rebel tamed as one who, stripped of all the trappings of class and privilege which may have blinded an audience to his true disposition, is dismissed as too shameful to mention."[36] By the time the flames rise, he has "genuinely traumatized" the women, a circumstance signaled in the narration scene by Martinpelto's disheveled cardigan and hair.[37] The relative blandness of the clothing robs the character of whatever power the gowns of old may have bequeathed, and the action reads similarly. Rather than standing still at the beginning, pointing dramatically after the recently departed Don Giovanni (as many Donna Annas do, even post-1980), Martinpelto's character takes several steps in pursuit, stumbles backward, collapses to her knees, and eventually sits on the floor, face haunted. The memories flood in before she gets the chance to voice them, and when she does begin to sing, bewilderment and fear undercut all signs of returning strength (fig. 9.2a; DVD, 51:44; chap. 20, 0:14). Performed seated and comparatively still, the recitative becomes a study in introspection, so much so that the character seems to forget where she is: after singing of her escape, she starts awake at Don Ottavio's interjection (fig. 9.2b; DVD, 53:38; chap. 20, 2:08). For the aria she stands, but the number begins lyrically, with rounded rather than accented high notes and appoggiaturas from below adding a hint of lament ("Or sai chi l'o-*no*-re"). After the close-ups of the narrative, moreover, wide camera framing distances the character from the viewer, and it also reveals her isolation from Don Ottavio, who sits well apart in his own pool of light (fig. 9.2c). Though surges of anger follow (complete with appoggiaturas

FIGURE 9.2 Donna Anna (Hillevi Martinpelto) and Don Ottavio (John Mark Ainsley), "Don Ottavio, son morta! . . . Or sai chi l'onore." *Don Giovanni*, dir. Warner 1995. DVD Kultur D2032, 2003.

a. The narrative begins

b. Donna Anna escapes

c. "Or sai chi l'onore"

from above), the performance never truly rages, and by the end the camera is picturing devastated expressions in close-up again. The aria becomes more supplication than demand, and the exit is an exhausted retreat.

Don Ottavio only makes things worse. John Mark Ainsley's character listens attentively and with visible concern, but he also turns away at times, as if too overwhelmed himself to be supportive (compare fig. 9.2b with fig. 9.2c). Elsewhere the fiancé demonstrates greater fortitude: Grobe's character evinces only compassion, and Luigi Alva's draws his sword and dashes back and forth searching for the criminal.[38] Yet a fair number exhibit shock or self-absorption, either of which leaves Donna Anna alone in her misery. Still others are actively unsympathetic, turning away repeatedly or behaving like investigators. In Sellars's South Bronx scenario, Carroll Freeman's Don Ottavio is a policeman and jealous boyfriend who responds to Donna Anna's narrative by trying to force himself on her. Her remembered escape from Don Giovanni becomes at once a real-time escape from Don Ottavio, and she leaves in fury—at him as much as at her attacker. Worse still is Benjamin Bruns's Don Ottavio in the Thilo Reinhardt production of 2011, one

---

Don Ottavio's Sigh

In Edward Dent's day, the reactions of Don Ottavio to Donna Anna's narrative were taken as further proof of haplessness. "Ottavio's little interjections of anxiety and relief during the long story of her encounter with Don Giovanni and his attempted rape of her become positively comical, and it is noticeable that a modern English audience almost invariably receives them with ribald laughter" (*Mozart's Operas*, 2nd ed., 161). Though one still hears giggles in the age of video, more often Don Ottavio looks self-absorbed rather than weak, preoccupied with his own honor as much as his fiancée's. The reveal comes when he hears how she managed to escape her attacker, twisting herself free. His reply is stilted at best, "Ohimè respiro!" (*Ohimè, I breathe again!*), and it is timed to a deceptive harmonic resolution (E^7–F) that many performances treat as a turn to introspection. While the statement is not marked as an aside, orchestras often pause before introducing the F major chord as quietly as possible, and singers enter sotto voce while looking off into space. From the Salzburg Festival in 1954 to the Royal Opera House in 2008, otherwise sympathetic Don Ottavios appear at this moment to be comforting themselves rather than their beloveds.

of several relocated into a world of twentieth-century fascism. A military officer in jackboots, he forces Mandy Fredrich's Donna Anna to testify before a crowd of priests, soldiers, and mourners taking part in the Commendatore's funeral procession.[39] Her sincerity makes him look a fool, but this does not diminish her isolation: defendant rather than confidant, she finds herself in the typical position of rape victims, pleading her case before male authority.[40] Telling the truth brings no justice, only further humiliation.

Torn

In the video record, Donna Annas who do not tell the truth—or not the whole truth—make their debut in the late 1970s. Both Peter Hall and Joseph Losey, directors respectively of a 1977 Glyndebourne production and the 1979 Gaumont feature film, decided that her version of events did not add up. "If an unexpected midnight marauder had come in through the window or made his way up the stairs and in by the door," opined Hall, "then Anna surely would have screamed a lot earlier than she did. I've an idea that Anna for the moment at least is besotted with this mysterious figure."[41] Losey added historical context: "If Donna Anna says that she is expecting someone at two o'clock in the morning and leaves her door open, that's more than a little strange for someone of that period. It must have been a previously arranged assignation."[42] Neither performance makes too much of these conclusions, but at Glyndebourne there are hints that Donna Anna pursues Don Giovanni out of awakened interest, and in the film there is a brief prologue (screened during the overture) in which the two characters exchange suggestive glances. Subsequent directors introduce more explicit signs of divided loyalties. In the late-1980s productions of Franco Zeffirelli and Michael Hampe, for example, Donna Anna submits willingly to a kiss in the introduction, scrambling away only when she hears the Commendatore coming. In Hampe's version, the character also turns away from Don Ottavio at the end of "Or sai chi l'onore," looking back toward the opening through which Don Giovanni has exited.[43] Passages in the narration take on double meanings as well. When they reach the words describing Donna Anna's escape (here in the Italian)—

a forza	by dint of
di svincolarmi, torcermi e piegarmi	freeing myself, wriggling and
	contorting myself
da lui mi sciolsi.	I released myself from him.[44]

—singers such as Grümmer and Martinpelto convey relief and exhaustion, and Teresa Stich-Randall looks victorious. In later productions, by contrast, the "wriggling and contorting" can become erotic and the escape a matter for regret or mixed feelings, depending on how mendacious Donna Anna is made out to be. In Martin Kušej's version of 2006, Christine Schäfer's character appears genuinely torn, while in Matthias Langhoff's of 1991, Marilyn Mims's is intent on deception from the start.

The last two examples are in modern dress, which frequently serves to connect Donna Anna's actions to broader social ills. In Bieito's underworld Barcelona, where everyone is dissolute and cheating a way of life, Regina Schörg's Donna Anna lusts openly after Don Giovanni and dupes Don Ottavio with a patently made-up story. Struggling to survive on the margins, she is motivated by no higher principles than self-interest and jealousy. Different pressures come to bear in more upscale environs, such as the urban penthouse created by Kušej and his designer, Martin Zehetgruber—"a world . . . that is sleek and cold, that is stamped above all by the omnipotence of sex and 'selling with sex,'" in the director's words.[45] Supernumeraries dressed as lingerie models embody the rampant commodification of women, and the ambivalence portrayed by Schäfer seems all too familiar, the coping mechanism of a socially entitled woman who finds herself the object of still more entitled men. A similarly bleak prospect fills the elegant hotellike surroundings of a 2006 production for Zurich directed by Sven-Eric Bechtolf and designed by Rolf Glittenberg.[46] Once again there are supernumeraries, and as Eva Moi's Donna Anna remembers her encounter with Don Giovanni—which has her sliding to the floor in abandon—a troupe of well-dressed couples enacts the ubiquity and violence of infidelity. Women strangle men and vice versa, their mounting bodies an index of moral turpitude among the affluent.

Other productions implicate Donna Anna in family traumas, echoing the psychoanalytic commentaries discussed above. For the 1998 festival at Aix-en-Provence, Peter Brook opted for middle-class clothes and an abstract space (designed by Tom Pye), reminiscent of Deborah Warner's production.[47] Yet Alexandra Deshorties's Donna Anna experiences a different kind of horror from Martinpelto's, at least on video, where her act 1 narrative is undercut by a rare cinematic intervention, a slow-motion flashback to "what really happened" in the introduction—namely, an encounter with as much embracing as fighting, and a Don Giovanni who wears no mask (DVD, 0:48:26; chap. 20, 1:05). Donna Anna knows who he is all along, and when she sings later of breaking free, the flashback shows the Commendatore's murder. Desire, liberation, and guilt jumble together in her mind, and the rest of the scene affords no relief. Conducted with exceptional swiftness by Daniel Harding, it leaves the tortured daughter weeping.

A dozen years later at the same festival, Dmitri Tcherniakov conveyed a similar message without the flashbacks. In the library of a family mansion, the Commendatore discovers Donna Anna having sex with Don Giovanni, and when she intervenes to protect her lover, her father is inadvertently killed. The narrative later in act 1 is frankly neurotic: driven again by lively conducting and grainy orchestral sound, this time courtesy of Louis Langrée and the Freiburg Baroque Orchestra, Marlis Petersen's Donna Anna undresses Don Ottavio as if to reenact the sex scene with him, only to push him off and retreat into despair.[48] Bechtolf offers a similar scenario in a 2014 Salzburg production that echoes his earlier collaboration with Glittenberg (the set is now a full-on art deco hotel) but with new Freudian overtones. Lenneke Ruiten's Donna Anna brandishes a knife during her own partfurious, part-sensuous encounter with Don Giovanni, only to be swung around by the libertine so that she stabs the Commendatore herself. Having literally killed the father, the character delivers her later narrative with simulated sex and threats to take the same knife to Don Ottavio—only now she suffers in slow motion, thanks to the measured pace of the Vienna Philharmonic, led by Christoph Eschenbach. From whirling confusion, the contest of desire and guilt becomes a drawn-out agony.[49]

As a final proof of family dysfunction, some productions add ominous backstories. Cast as another jackbooted fascist, the Commendatore in Bechtolf's Salzburg version is one of several fathers who recall the self-interested Don Ottavios discussed above. Prizing their honor above their daughter's safety, they react to her cries with anger or disgust—in Bieito's production, the character smashes a car window with a baseball bat, causing Donna Anna to flinch and suggesting that they have a history of violence. In the 2007 Netherlands production by Jossi Wieler and Sergio Morabito, a symbolic gesture gets at the same idea; Donna Anna tells her story sitting on a bed next to the dead Commendatore, whose leg rises up like an erection. Whatever happened with Don Giovanni (the staging is deliberately vague), the memory is intertwined with one of paternal abuse. In such scenarios, Donna Anna is not simply attracted to the wrong man, nor frustrated by her fiancé's coldness, nor beset by unfair expectations of women. Her behavior is rooted in psychoanalysis's classic primal trauma.

Humanized?

Their attraction to Don Giovanni notwithstanding, none of these conflicted Donna Annas has the mythic stature of Hoffmann's heroine, passionate counterpart to the demonic libertine. On the contrary, and despite all the family archetypes, the emphasis on social and personal circumstance tends to demythologize the character, portraying her motivations as familiar and mundane. More than one director speaks of humanization, the recurring trope of Mozart criticism that is taken here to signal an improvement on older, clumsier modes of interpretation. "Now that opera has been freed from the horrors of opera tradition," cracked Peter Brook in 2002, sympathetic insight assumes its proper role: "We tried to do it [*Don Giovanni*] with very, very open singers who were as interested in the humanity of their characters as in the external sides of characters in opera, [who] really felt that these were human beings."[50] The wayward desires of his production's Donna Anna, and her confusion of fury and remorse, speak not to a grand destiny but to ordinary experience. Kasper Holten applies similar criteria to his Royal Opera House production of 2014, in which Donna Anna sleeps willingly with Don Giovanni even after the necessity of lying to Don Ottavio has left her visibly upset: "How often don't people do stupid things, they wreck their own lives, they drink although they know they shouldn't, they fuck around although they know it will wreck their family life, you know, it happens all the time, it's human, so it's not right, it's not logic, but Anna here does it anyway."[51] Realism demands that the characters make bad decisions, no matter how self-destructive. Having enmeshed herself in a comparable quandary, the Donna Anna of Claus Guth's 2008 production for Salzburg repeatedly threatens suicide.[52]

Humanism can take on political dimensions as well. By the end of Holten's production, Donna Anna has rejected both Don Ottavio and Don Giovanni, even though the latter makes an unscheduled appearance during her final aria, "Non mi dir, bell'idol mio" (as he does also in Guth's production). Nonetheless, the notorious coloratura signals her determination to move beyond her mistakes and forge a new life, giving the lie to the passivity sometimes associated with the opera's women:

> I thought it was quite important, actually, in the piece to even up the balance between men and women. Some people even accused me of being misogynist because some of the women are more active players. . . . I felt

it was actually the opposite . . . traditionally, as is often done, portraying the women as innocent victims, who just don't understand anything of what's going on, and him [Don Giovanni] being able to completely manipulate them, you kind of reduce them to . . . innocent victims, and I thought it's much more interesting to have a more complex world, where the women he is with also have their own agendas.[53]

Holten's reasoning has its own history in production and criticism. As early as 1984, the director Ruth Berghaus used *Don Giovanni* to critique, in Tom Sutcliffe's account, "the conventional equation of sex, politics and religion that always puts women down. Her central point was that for Mozart's women the Don was more liberator than exploiter. . . . Berghaus's staging freed the females in the story from permanent victim status."[54] To judge from descriptions, her production had little in common with those I have been discussing, eschewing realism and anything else that might induce "sentimental identification with the events shown."[55] Still, it figured sex with Don Giovanni as opening new horizons for the women, a reversal that echoes into Bieito's much later urban realism, as read by Sarah Wright: "There is no rape: sex is graphic, public or furtive, but takes place between consenting adults. Bieito's females are stricken with love-sickness, and with an obsessive desire for unattainable ideals, yet they are also earthy pleasure-seeking sexual beings who lament loudly if treated badly by a sexual partner."[56] Whatever her motivation, Donna Anna has every right to pursue sexual satisfaction.

After all the celebration of Don Giovanni's sexual pursuits, turning the tables like this is undeniably attractive. A sexually emancipated Donna Anna rebukes the paternalist demand that she "get over it" and embrace her fiancé, just as a wise or manipulative Zerlina eludes the voyeurism so often aimed in her direction (see chap. 4). When Don Ottavio or Masetto comes across as suspicious or judgmental, the women's infidelity may even look like brave resistance to male hegemony. That said, barring more radical revisions than directors have thus far been willing to make, having the women choose sex may not empower them so much as expose them to new hegemonies. For Zerlina, whatever freedom she finds in an affair with Don Giovanni comes at the expense of further subjugation; the new master may be more cosmopolitan than the old, but he is still a master. An unfaithful Donna Anna is subject to the same logic, and thanks to the narration scene she also confirms another enduring stereotype about rape victims: that they invent their stories in order to protect their reputations. It is in this respect that productions such as Holten's can seem misogynistic, however much he and others want to represent women as authors of their own sexual destinies. There is simply no way to turn Donna Anna into Don Giovanni's lover, or even his admirer, without also making her a liar. When she embroiders her recitative with sensuous writhing or sidelong glances, the performance risks endorsing a long history of casting doubt on women accusers.

Equally double-edged is the association of concealed desires with humanist insight. "But to think that Donna Anna was somehow burned by the erotic flame of Don Giovanni," wrote Massimo Mila in 1988, ". . . that in persecuting her rapist she wants in a certain sense to extinguish and punish in herself that moment of weakness which was perhaps going to overpower her, this is not forbidden: the music supports it, and certain situations in the drama seem almost to suggest it. The complexity of the character has everything to gain."[57] It would be more accurate to say that imbuing Donna Anna with conflicting impulses substitutes one kind of complexity for another. At once furious and deathly afraid, affectionate and wracked by grief, the character played by Elisabeth Grümmer or Hillevi Martinpelto can hardly be called simple. She may not change her mind like Zerlina or Donna Elvira, but she is no less buffeted by intense and contradictory feelings. To cast her as unfaithful, or as a victim of peer pressure or childhood abuse, is to suggest that the attendant traumas are somehow "deeper" than those triggered by violence. No doubt they better satisfy conventional expectations of women characters; a sensual, deceptive Donna Anna is more amenable to Hoffmannesque fantasies of possession than her angrier counterparts. But she is by no means more psychologically plausible. Ironically, the long quest to get at Donna Anna's inner life has too often obscured the painful and complex traumas brought on by sexual assault.

CHAPTER TEN

Libertines Punished

In an opera full of startling juxtapositions, the concluding scenes of *Don Giovanni* are in a class by themselves. As the libertine confronts divine justice, his sidekick blubbers with mundane fright, and after hell itself opens and closes, a denouement has the others asking, "What's next?" Convents and new masters and pastoral bliss, they say, and then everyone spouts a platitude:

Questo è il fin di chi fa mal:	This is the end of evildoers:
e de' perfidi la morte	and the death of the wicked
alla vita è sempre ugual.	is always the same as the life.[1]

This is a stretch even for opera buffa. Though also equipped with improbable happy endings and morals, neither *The Marriage of Figaro* nor *Così fan tutte* asks performers to negotiate between supernatural and natural, glowering D minor and peppy D major, earthshaking threats and glib banalities. More than implausible, the end of *Don Giovanni* seems all but unperformable.

The Romantics agreed and stopped with Don Giovanni's death. Why spoil his immolation with the caviling of mortals? Unless pledged to the Vienna version, however, most audio and video recordings retain everything—and with it, generations of attempts to bridge moral parable and social drama. Performances from the 1950s and 1960s illustrate the Romantics' long reach, treating the statue scene as a breakthrough into the sublime and the aftermath as a prayerful reaction: if the mortals cannot be ignored, at least they can be made solemn. In subsequent decades, doubts about Don Giovanni diminish his fall and push the bystanders toward outrage or myth-puncturing irony, and after 1990 this trend was joined by efforts to expunge the supernatural and subject the libertine to secular punishments such as murder, ostracization, or psychological breakdown. And not just him. Some of the most unified finales forge conflagration and denouement into one long punishment of everyone, an indictment of society itself. Breakthrough turns to breakdown, leaving only guilt and the weary knowledge that it can all happen again. The survivors sing their platitude about themselves, as if they, too, had no more to look forward to than a wicked life followed by a wicked death.

Grand Romantic Opera

For a Romantic apotheosis, no recording quite matches the *Don Giovanni* made at the Salzburg Festival in 1954. The festival commissioned a vast Renaissance cityscape from Clemens Holzmeister, before whose towering façades pace richly costumed stars, accompanied by the sonorous Vienna Philharmonic under Wilhelm Furtwängler. The recording itself is no less grand, a color film directed by the cinema veteran Paul Czinner using material from multiple sessions staged without audience.[2] Though not without quirks, particularly in the soundtrack, the result far outstrips the more typical black-and-white broadcasts of the day, and its 1994 video release has kept an unabashedly larger-than-life *Don Giovanni* very much in circulation.[3]

The tenor of the final scenes is announced by Deszö Ernster's statue, who looks and acts the part of a solemn judge from beyond. Clad and made up all in gray, he steps through an imposing double door and then remains still, moving only to sing and eventually to accept Don Giovanni's hand in pledge (fig. 10.1a). The cameras maintain a respectful distance, and when he is pictured alone the focus blurs slightly, further distinguishing him from the bright colors and sharp images of the human characters. Or rather, character: crouched off to one side, Otto Edelmann's Leporello appears only in three fleeting cutaways, robbing the scene of comic relief and any sense of human scale. Instead, it is the shock, fear, and courage of Cesare Siepi's Don Giovanni that fill the screen as he casts frightened glances and worries at his hair but also attacks the statue with a knife and repeatedly draws himself up in defiance, jaw thrust forward (figs. 10.1b and c). The pose is heroic, but the handshake breaks him all the same, and after

FIGURE 10.1 Don Giovanni (Cesare Siepi) and the statue (Deszö Ernster). *Don Giovanni*, dir. Graf 1954. DVD Deutsche Grammophon 0730199, [1994] 2001.

a. The statue sings

b. Don Giovanni in doubt

c. Don Giovanni defiant

writhing at the statue's feet, he runs. Earlier in the opera, Siepi's dashes across the cityscape evoked the Romantics' equation of Don Giovanni with vitality, but this time he gets only as far as the waiting Furies, who come amidst smoke and flickering red lights to drag him off into the deepest of Holzmeister's archways. However effective with the humans, vitality alone cannot overcome the punishing flames or their grave herald.

Gravity marks the musical performance as well, beginning with an ear-splitting, soundtrack-distorting rendition of the scene's massive opening chords (DVD, 2:41:10; chap. 49, 0:02). The same grandeur carries through the ensuing andante, in which the statue invites Don Giovanni to dinner; through a più stretto, in which he demands that the libertine repent; and finally into an allegro, in which the Furies thunder and Don Giovanni descends. All three sections are marked alla breve, suggesting two half-note beats per measure, but Furtwängler's andante is clearly in four, beginning around 60 bpm and fluctuating downward at several points. At this speed the statue's statements, which track the half-note pulse, take on an epic quality, rolling out so slowly that Ernster struggles to sustain them. The accompanying dotted quarter–eighth figure, traditionally associated with the statue's tread, likewise sounds more like the stone itself, played without lift and anchoring the music in place. The other characters are no less affected, though their lines feature shorter note values. Even the triplets sung by Leporello, setting his nervous asides, become in this context an ominous rumbling from beneath (e.g., DVD, 2:43:41; chap. 49, 2:34).

The following sections bring little change. Though the quarter-note pulse circles around 100 bpm in the più stretto, the accentuation shifts toward the half note, such that the pace may actually seem to slow, to an alla breve of 50 bpm (DVD, 2:47:12; chap. 49, 6:05). And while the allegro reaches approximately 72 bpm to the half note, even in this era the norm lies in the upper 80s and 90s, and neither the Philharmonic nor the singers show any signs of lightening up (DVD, 2:48:33; chap. 49, 7:26). On the contrary, where many Don Giovannis and Leporellos lapse into shouting at the end, Siepi and Edelmann continue in full voice. With their twin basses projecting exceptional power (more often one of the two characters is cast as a baritone), they register shock and awe but no panic.

Commenting on a radio broadcast of the 1954 production, itself reissued on CD in 1986, the critic Alan Blyth remarked, "This historic performance presents *Don Giovanni* very much as Grand Romantic Opera."[4] The description rings equally true of the film, whose stately visuals, massive sound, and epic time scale seem almost Wagnerian. More specifically, its seriousness is a properly Romantic answer to the restless energies of Siepi's Don Giovanni. If vitality powers his defiance, the response is as titanic as Hermann Abert could ever have imagined when he wrote that "one demon can only be conquered by another" (see chap. 3). A superman Don Giovanni cannot be hurried off the stage; he is too strong, conceptually and musically, to fall prey to accelerating tempos or even a shouting chorus. Siepi's libertine is not consumed so much as crushed, forced from the world by rhythmic and sonic weight.

Another commentator quoted in earlier chapters, Heinrich Bulthaupt, made the Romantic case against the ensuing denouement. More than just dragging audiences from "the spiritual heights of the apparition . . . to the flat ground of the conventional," it undermines a heroic ideal: "Someone whom heaven honors with such a direct intervention into his existence, whom eternal justice visits in person—he must be an exceptional man. Divinity ennobles its creations even with its judgments . . . For such a blessed one to be sought posthumously by the guardians of earthly law with spears and clubs, and to be apostrophized with a Philistine moral—what a debasement, what a humiliation!"[5] Furtwängler and the stage director at Salzburg, Herbert Graf, might well have agreed: having made their own protagonist "exceptional," they could have followed nineteenth-century tradition and left audiences in thrall to his death. Instead they scripted a response, depicting those left behind as reflective rather than vindictive. Like the statue scene, the denouement also is divided into three parts: an allegro assai, in which Leporello recounts the events; a larghetto, in which the characters look to the future, led by Don Ottavio and Donna Anna in an extended duet; and a presto, in which all sing the moral together. Furtwängler takes moderate tempos throughout, even in the presto, which is slower than most at roughly 130 bpm. The emotions are subdued, their development controlled. Edelmann delivers Leporello's narrative seriously, taking the statue's place in front of the double door and showing no lingering terror or bewilderment (DVD, 2:50:31; chap. 50, 0:30). His listeners, too, rather than rushing about looking for the missing hero, attend quietly or kneel to pray (they include a crowd of supernumeraries, magnifying the impact of the catastrophe). The larghetto begins similarly, the noble lovers standing still and avoiding histrionics as well as some of Mozart's coloratura; as in many recordings of the period, the duet is trimmed, saving time and reducing the chance that vocalism will overshadow sentiment (DVD, 2:51:51; chap. 50, 1:50). The other characters are more dynamic, especially Zerlina (Erna Berger), who resumes her innocent perkiness; but even she can enliven things only so far. For the moral, finally, the principals fall into a line, gazing into space as they preach not to a live audience, nor even to the camera as on some videos, but to the world at large. The words may still sound philistine, to the ears of a Romantic, but the performance does everything it can to prevent trivialization. These survivors know an epoch-making event when they see one, and they react with all due humility.

Romantic but Not So Grand

Romanticism did not soon loosen its grip. Until at least 2000, most video-recorded *Don Giovanni*s include elements of the Salzburg conclusion: immobile statues, undaunted heroes, smoke and fire, pensive survivors in a line. The setting remains grand, especially on big stages; at the Salzburg Festival in the opera's bicentennial year of 1987, Samuel Ramey and Paata Burchuladze collide before the only backdrop that could possibly outdo Holzmeister's city: a cosmos of planets and stars.[6] And while no one sounds quite like Furtwängler, not even Herbert von Karajan with the same orchestra in 1987, the music retains plenty of grandeur in its own right. Leaving aside period-instrument versions, a good 80 percent of both audio and video recordings through 2000 feature statue scenes in which the andante unfolds in four beats per measure at 65–75 bpm, the più stretto in two at the same speed or a little faster, and the allegro in two at 80–90 bpm. With the tempos come the requisite vocal heroics, orchestral sonorousness, and (mostly) heavy accentuation. In the denouement, the opening remains similarly serious and the duet heartfelt; only the presto may achieve real incandescence at 150 bpm or more. Musical performances still aim to split the world open, then to rid the aftermath of any undue frivolity. Video directors contribute something as well, editing for spectacular effect and focusing on the protagonist at the expense of the others, in some cases even the avenger. A statue does not make for great video, of course, but picturing Don Giovanni even while his nemesis is singing adds a further Romantic twist. The ghost may determine the outcome, but the drama is all about the hero.

Romantic conventions linger even into the revisionist interpretations that began to appear in the 1970s, with complex and sometimes contradictory results. A good example is the 1977 Glyndebourne production mentioned in the last chapter, directed by Peter Hall. Though he updates the story to a bona fide Romantic setting, the British Regency of the early nineteenth century, Hall has no visions of superman: "I was looking for someone of considerable personal warmth because I thought it was only with this that I could risk the chill, cruel self-centredness of Giovanni. Only very sympathetic rich personalities can portray outrageous shits on the stage without apologizing for them."[7] And indeed, while possessed of formidable charisma, Benjamin Luxon's character does things that Siepi's never would, such as grin before killing the Commendatore and lick his lips while appraising Zerlina. His death is accordingly unheroic. Behaving more like a living being than a stone, Pierre Thau's statue drags Don Giovanni upstage toward an illuminated cross, diminishing the libertine on two counts: not only is he attacked in his person, but he is defeated by plain Christian morality rather than cosmic forces. The costs of his actions are evident as well, in a Leporello whose efforts to intervene are violently rebuffed. Stafford Dean's face becomes an emblem of luckless humanity, visible even against the rising flames (courtesy of a dissolve by the video director, Dave Heather).

On the other hand, in sonic terms the scene remains as grand as ever. To be sure, under Bernard Haitink's leadership the statue's tread has some lift, and an andante tempo of about 65 bpm allows Thau to sustain the statue's phrases. All the same, the tempos remain slow enough, and the sonorities thick enough, to ensure that the struggle remains in slow motion. The music aggrandizes as the staging downsizes, a contradiction found also in the Joseph Losey film of 1979, in which a Marxist critique of the plot coexists with a lush reading of the score, conducted by Lorin Maazel. If the world no longer cracks open, its deliberate turn becomes a force in its own right, confirming the libertine's strength even as his faults become more visible.[8]

Romanticism recedes a little further when the musical confrontation grows more dynamic. In 1960, in a studio production by the NBC Opera Theater, the conductor Peter Herman Adler apparently decided that American television audiences would prefer a fight to a symbolic confrontation.[9] Beginning at about 50 bpm (beating the half note), the statue scene accelerates from there, and though the staging remains static (two men at opposite ends of a banquet table), the dialogue is transformed. Only six years after playing a demigod in Salzburg, Siepi becomes a combative Don Giovanni striving against the equally active statue of John Macurdy. Something similar happened at La Scala in 1987, in a production featuring a prominent Don Giovanni of the new, no-longer-so-charming school, the baritone Thomas Allen. While not abandoning Siepi's warmth, Allen added the "cruel self-centredness" of Peter Hall's conception (Allen debuted the role in a later iteration of the Glyndebourne production) and a creepiness recalling the classic literary seducers of Pierre Choderlos de Laclos (*Les liaisons dangereuses*) or Søren Kierkgaard (*The Seducer's Diary*). As the singer wrote: "Stillness—I now lay great store on the power of stillness. An almost hypnotic concentration—such as those moments when Giovanni, at his own party, sizes up Zerlina as his next prey."[10] Under duress, Allen's version of the character exchanges stillness for bravado, but signs of desperation diminish the scope of his fall. In a 1992 Royal Opera House production, a musical performance led by Haitink preserves the old deliberate heroism, but Johannes Schaaf's direction has the libertine clutching pathetically at the statue's cloak as he dies.[11] Five years earlier at La Scala, a similar staging by Giorgio Strehler—this time, Don Giovanni clutches after Leporello—found compelling musical reinforcement.[12] Riccardo Muti launches the andante briskly, and the ensuing dialogue is full of urgent back-and-forth. The energy builds to a savage climax: in an allegro as swift as any period-instrument version (ca. 105 bpm), the every-other-measure *forte*s evoke flashes of lightning, and the dotted rhythms of the chorus of Furies become the sound of a ticking clock. Time replaces mass as the agent of punishment, sweeping Don Giovanni away without ceremony or grandeur.

The death has consequences for the aftermath. Whereas Graf and Furtwängler solemnified the final scene, Hall and Haitink, and later Schaaf and Haitink, take an ironic turn. The duet for Donna Anna and Don Ottavio remains lyrical and serious, but the rest of the characters relax: Donna Elvira declares resignedly for the convent, and the lower classes chuckle over their own plans. For the moral, Hall's singers join hands and address the camera, which eventually reverses angle to look past them into the house (at Glyndebourne as at Salzburg, the video records a performance mounted specially for cameras). No more lining up to deliver a serious message; rather, the platitudinous words and presto tempo drive a leap into pure theatricality. Schaaf's ending is similar except that the leap ends in social artifice, a tableau of acquaintances chatting politely over tea.

Strehler has the cast sing the presto in front of the curtain, suggesting, again, that the moral lies outside the story proper. Images of Muti conducting, floating above the characters in half-dissolve, further remind video viewers that the entire spectacle is after all an opera performance. That said, several earlier scenes have taken place before the curtain, and with wreckage from the banquet visible behind the singers, the space seems more integral to the drama than one might otherwise assume. Muti, who hovers in similar fashion over the maskers' prayer in act 1, affirms by his presence

Though recorded only a few years apart, the act 2 finales led by Riccardo Muti (*Don Giovanni*, dir. Strehler 1987) and Craig Smith (*Don Giovanni*, dir. Sellars 1990) illustrate the remarkable variability of the opera's most iconic passage. Muti and the La Scala orchestra establish a tense atmosphere in the opening measures of the andante, articulating the treading rhythm sharply and at a tempo approaching 80 bpm to the quarter note. This allows Sergej Koptchak, playing the statue, to sing in complete and directed phrases rather than the isolated fragments of predecessors such as Deszö Ernster or Pierre Thau, who have to pause mid-sentence at critical moments (e.g., "Dammi la mano"—[*breath*]—"in pegno!" [Give me your hand—in pledge]; see *Don Giovanni*, dir. Furtwängler 1954, DVD, 2:46:55; chap. 49, 5:48). Koptchak instead builds an inexorable momentum, aided by Muti's use of rallentando, first in the measures just before the statue invites Don Giovanni to dinner (DVD act 2, 1:10:51; chap. 14, 10:22), then again at the end of the andante (DVD act 2, 1:12:30, chap. 14, 12:03). The first instance prepares an exhilarating return to tempo, and the second, an even more exhilarating leap into the più stretto, which at around 90 bpm is as fast as the concluding allegro in some recordings. Paradoxically, though Koptchak sings from offstage, his conversation with Thomas Allen also brims with give-and-take, a real struggle ensuing between duple and triple rhythmic subdivisions, pushing forward and holding back, clarity and chaos (DVD act 2, 1:10:07; chap. 14, 9:38). Smith and the Vienna Symphony set a very different tone, starting with the opening chords, which are not the usual scarifying blasts but comparatively lyrical and sonically well-balanced intonings that swell to full volume (DVD disc 2, 1:20:48; chap. 11, 0:01). The ensuing andante continues in the same vein, more somber than urgent or momentous. Though Smith's tempo is slow, ranging from around 66 bpm to the quarter note all the way down to Furtwängler's 60 bpm, the singing is intimate rather than grand: where other statues thunder, James Patterson's laments, and where other Don Giovannis spit out their retorts—"*Parla dunque . . . Parla, parla*" (Speak then . . . speak, speak; Gronda, *Il Don Giovanni*, 76-77; Stivender and Webb, "*Don Giovanni*," 371–72)—Eugene Perry's shapes them carefully and goes light on the opening consonants (DVD disc 2, 1:23:44; chap. 11, 2:57). Theirs is a prolonged and emotional dialogue, preparation for a tragic version of the death to come.

the musical weight of the conclusion: this is no frivolous stretto, but a necessary punctuation. The characters, too, seem anything but ironic; rather than chuckle or look arch, they address the audience as if they meant it, as if the opera really did prove that wicked people meet wicked ends. Theirs are serious reflections, just as the libertine's death, its grandeur faded, is more sobering than enthralling. *Don Giovanni* is edging from the Romantic toward the didactic, a sign of things to come.

Social Dysfunction (I)

The decades following Hall's and Strehler's productions brought many more wicked Don Giovannis to stage and video, from the middle-class pervert of Deborah Warner's version, to the vicious playboys of Martin Kušej's and Sven-Eric Bechtolf's, to the glowering Oedipus of Sergio Morabito's and Jossi Wieler's—to mention only examples that have come up in previous chapters. Their degeneracy is emphasized by a turn toward more graphic sex and violence onstage, toward naturalism more generally, and toward musical grit. These are all updated stagings, but their traditional counterparts exhibit similar tendencies. When the Met revived its Zeffirelli production in 2000, a decade after the premiere, it kept his eighteenth-century designs but introduced new stage direction by Stephen Lawless, and a more emphatic if still deliberate musical performance led by the original conductor, James Levine. Few examples better illustrate the changing times: in the same ornate surroundings, Samuel Ramey in 1990 comes across as a mostly gallant deceiver, but Bryn Terfel in 2000 as a depraved murderer and sexual predator. The words of the conductor Craig Smith, describing Peter Sellars's real-crime South Bronx version, could apply equally well to the Met revival: "There's been a kind of grisly old tradition of making Don Giovanni a fun sort of guy. We feel that he's not."[13]

The 2000 performance also foregrounds the gender and class inequities that benefit the protagonist. Watching the character played by Cesare Siepi, it is easy to forget the leverage he enjoys as a man and a Don: the very picture of chivalry, he seems to rely on looks and charm rather than force, acting the gentleman even with the most socially vulnerable of the women, Zerlina. The character as played by Terfel or Luxon, or by Ruggero Raimondi in the Losey film, is more clearly master of his realm, freely deploying the wealth, authority, and monopoly on violence that come with being an aristocrat. Many updated Don Giovannis behave similarly, exercising whatever powers they command in the twentieth- and twenty-first-century hierarchies represented onstage, be they economic, familial, criminal, or military. Beyond personal wickedness, they illustrate the corruptions of entitlement, and their story becomes that of "a monstrous world and its deliverance," to recall a description quoted in the introduction.

At the same time, and as we saw in the last chapter, making social conditions part of the story can have unpredictable results. The most obvious case may be Sellars's production, whose social criticism depends on an equation between status and race.[14] *Don Giovanni* had evoked racial hierarchies before: as James Parakilas argues, the frequent pairing of Italian singers such as Siepi or Ezio Pinza with Northern or Eastern European sopranos "would have quietly suggested a scenario of fair, innocent Nordic womanhood being seduced by dark, lustful Latin manhood," while the casting of Black women singers amidst otherwise all-white ensembles, beginning with Leontyne Price as Donna Anna, "would dramatize the social isolation of the character."[15] Making the implicit explicit, Sellars turns the entire drama into a metaphor of U.S. race relations, depicting an inner city in which the lower-class characters are Black or Asian, the upper-class ones nearly all white, and the whole environment rife with crime. The setting abounds in urban clichés, none more troubling than the character of Don Giovanni, a Black drug dealer who upends the social order by way of money, profligacy, and violence. One could classify him alongside other updated protagonists, except that a Black male criminal onstage is never simply another bad guy, but a figure so redolent of white-supremacist constructions of blackness that the production risks appearing racist. The other characters cast as Black men, Masetto and Leporello, offer little corrective: the former literally beats Zerlina in the lead-up to "Batti batti, o bel Masetto," and the latter, though more conscience-stricken than the protagonist, is represented as family—the parts are played by the twin brothers Eugene Perry (Don Giovanni) and Herbert Perry (Leporello). Whatever the parallels between racial and class oppression, criminal Black characters reinforce invidious stereotypes in the same way that lying Donna Annas do.

The production also aspires to critique the stereotypes. Like an inversion of Holzmeister's ideal cityscape, George Tsypin's design offers a run-down city block, complete with graffiti and blinking construction barrier (fig. 10.2). Its realism is seconded on the video by the inclusion of documentary-style footage at beginning and end, dominated by shots of crumbling tenements and rat-infested vacant lots. The inner city is no thrilling danger zone, but a social and political calamity in which the lib-

FIGURE 10.2 George Tsypin, cityscape for *Don Giovanni*, directed by Peter Sellars. From *Destination Mozart: A Night at the Opera with Peter Sellars*, directed by Andrea Simon. Arcadia Pictures, 1990. DVD Kultur D1288.

ertine himself seems as much victim as perpetrator. He is twice seen stripping to his underwear, first when confronted in the act 1 finale, then again when condemned by the statue. The second instance has precedent in other productions (there is often undressing at the end, if not so much), but the spectacle of his naked, vulnerable body inevitably brings to mind lynching and other forms of violence perpetrated against Black men. In act 2 he is further confronted by racialized evidence of his wrongdoing, including a coffin, reminder of the white man he may or may not have murdered,[16] and a young blond girl, embodiment of white female innocence. It is she who eventually grasps his hand, suggesting that his real nemesis is not some make-believe ghost, but the all-too-real paranoia of white society about Black male sexual aggression.

Along similar lines, despite horror-movie theatrics involving a banquet from McDonald's and a ghost glowing radioactive green, the focus at the end is on the libertine's humanity. Together with mostly naturalistic acting and exceptionally close-up filming, the setting figures the characters as real people, an impression that may be heightened for English-speaking viewers by the subtitles, written in a slangy colloquial by Sellars himself. Don Giovanni's parting gasps, translated literally in EMI's video of the Salzburg film ("My soul is torn in agony / My body is in torture! / What torment! What madness! / What hell! What terror!"), become simply, "Oh God! The madness!" The words express a concentrated terror whose visual correlate is wide-eyed shock; even if he will not repent, the character at least recognizes his sins. The music seems accordingly life-sized, eschewing urgency and grandeur alike in favor of an unhurried and seemingly heartfelt expressivity. Neither proud Mephistopheles nor snaky deceiver, the urban Don Giovanni dies as a feeling subject, a human pawn of social inequity.

Whether the production succeeds in exploding rather than perpetuating racist stereotypes is an open question. Viewers may be led to recognize a real-world system of oppression whose beneficiaries include the generally white, affluent institution of opera itself. On the other hand, they may equally be led to regard the protagonist's crimes as symptoms, rather than causes, of a social dysfunction imposed from without. As Naomi André shows, the latter view can push Black opera characters toward minstrel stereotypes, exempt from judgment so long as they furnish entertainment. When the hero of *Porgy and Bess* escapes punishment for committing murder and then goes on to lead an uplifting final chorus, "It is almost as if the artificially triumphant ending tells the audience that these poor black people should not suffer the consequences of their actions, because, after all, they can't really help themselves for not knowing better."[17] Sellars's criminal pays the price for his deeds, and the denouement offers no catharsis: Leporello looks distraught, while the rest of the survivors pray against a background of documentary footage. Nevertheless, the specter of systemic oppression raises the question of whether a Black Don Giovanni in these circumstances chooses libertinism or has it forced upon him.

Social Dysfunction (II)

Though not without their own contradictions, socially critical productions since Sellars's tend to evince less sympathy toward the protagonist—or anyone else, in some cases. The violent, drug- and alcohol-fueled cast of Calixto Bieito's version stumbles through a Barcelona only slightly less decrepit than Tsypin's New York, but they make a lot of the mess themselves, leaving a trail of litter to match Don Giovanni's trail of discarded lovers. If they are discards in their own right, losers in the neoliberal economy, they encounter nothing so fraught as a lynching or a sexual assault, and their behavior appears self-destructive at best. In tonier surroundings, the relative privilege of the characters makes their actions look worse. Amidst the immoral one-percenters of Kušej's and Bechtolf's productions, or the uniformed fascists of several others, Don Giovanni and occasionally the other men inflict visible suffering on a range of victims, including supernumeraries. Their degeneracy is highlighted by ever more severe punishments. In Sellars's ending, Don Giovanni slips away without further violence or violation, but since 2000 he has been confronted by the ghosts of his conquests; squeezed until his veins pop by a grotesquely decomposed Commendatore; and shot at the beginning of the opera so that he bleeds out gradually before toppling into the Commendatore's grave—watched by the bloody ghost himself.[18] Elsewhere he loses the honor of a supernatural intervention, collapsing from the stress or getting stabbed to death by the other characters. The libertine of Thilo Reinhardt's production, played by Josef Wagner, suffers all of it, perhaps to balance a nightmarish career that ends with him feasting on the bodies of his dead victims. They arise first, chasing him around the stage, then a zombie Commendatore helps maneuver him into a coffin, then a firing squad lets fly—all for a character who has been bleeding out since the swordfight at the beginning of the opera.[19]

Retribution feels even more savage in the era of musical edge. A few examples summon the old Romantic grandeur, notably Bechtolf's production for the Salzburg Festival, in which Christoph Eschenbach and the Vienna Philharmonic echo the deliberate pacing and sonorous boom of older renditions.[20] More typical, though, are the Salzburg Festival recordings of Kušej's production, or of Claus Guth's. Though leading the same orchestra as Eschenbach, Daniel Harding and Bertrand de Billy proceed more like Muti at La Scala, turning spectacle into struggle by means of quick tempos, sharp accents, and clipped delivery. Some years earlier Billy went even further in Bieito's blood-soaked finale, in which a still-living Commendatore climbs out of an automobile trunk and Don Giovanni cooks him an omelet before finishing him off (Don Giovanni does not die until the denouement—more on that later). Echoing Peter Hermann Adler in 1960, or more recent period-instrument versions, Billy's andante exceeds 50 bpm of alla breve, accelerating the insult-slinging into a frantic blur. Hard attacks from the Orchestra Academy of Barcelona's Gran Teatre del Liceu and raw exchanges between the characters continue to roil the atmosphere even as the tempos of later sections return to more moderate levels.

As for period-instrument versions, grit comes into its own in a scene that makes heavy use of timpani and brass, including a trio of trombones. In a Glyndebourne production directed by Jonathan Kent, the one with the exploding Don Giovanni (Gerald Finley), the musical texture experiences a comparable rupture. Crowning an aggressive performance by Vladimir Jurowski and the Orchestra of the Age of Enlightenment, the timpani and brass all but overwhelm the soundtrack at the end, tearing at the fabric of the score and substituting noise for music. Conducting the Balthasar-Neumann-Ensemble for a production by Philipp Himmelmann, Thomas Hengelbrock achieves a similar effect more incrementally. The andante begins at Muti's speed (and Jurowski's) of around 80 bpm to the quarter note, and the mood is gentle: heard from offstage, Mario Luperi's Commendatore sounds almost pleading, and the main sign of agitation comes from Erwin Schrott's Don Giovanni, who breaks occasionally from singing into shouting. But Schrott has played a domineering tough, and to prepare his grim demise (vanquished by his victims' ghosts), the performance gets steadily more insistent. Voices and orchestra grow harsher, and the pulse ratchets up, exceeding 100 bpm during the concluding sections and, in one delirious accelerando, nearing 120 bpm (BD, 2:40:50). If there is an early twenty-first-century normal for Don Giovanni's end, it is an unmajestic, unsentimental, sometimes uneven surge of sound that dispatches him to infamy.

All of which leaves his fellow characters in a sorry state. As in the 1970s and 1980s, the denouement exhibits more musical uniformity than the statue scene, with most performances moving from a moderate andante, to a slow and genuinely lyrical duet, to a quick and exciting presto. A few examples introduce cuts: both Sellars's and Kent's productions omit the duet, an alternative that originates in the sources of 1788, and Reinhardt's skips the entire andante. At any length, however, the passage tends toward a somberness that resembles neither the solemnity of the Romantics nor the irony of directors such as Peter Hall and Johannes Schaaf. In some instances, everyone looks sad or stunned, and the pairs of lovers remain separated and self-absorbed right to the end. A dictum of Kierkegaard's comes to mind: "The very secret of this opera is that its hero is also the force in the other characters. Don Giovanni's life is the life principle in them."[21] Jurowski has a more negative spin: "Everybody needed Don Giovanni, even to punish him or to get the debts paid from him, all because people were dependent upon him and his evil and yet incredibly strong and therefore fascinating energy."[22] Other productions are still more negative, using the denouement and especially the platitude to spread the evil around. Finding Don Giovanni alive, Bieito's characters tie him to a chair and take turns knifing him, with a reluctant Donna Elvira delivering the fatal blow. Lluis Pasqual's characters, denizens of 1940s Spain, come together instead to make a propaganda newsreel featuring images of fascist rallies and a slogan of Generalissimo Franco's, "El mundo entero al alcance de todos los espa-ñoles" (The whole world within the reach of all Spaniards). Naïve enough to attribute their deliverance to the regime, they miss the fact that Carlos Álvarez's Don Giovanni, himself a general in uniform, has returned from the dead to direct the newsreel. Murderers or collaborators, the victims are as guilty of injustice as the libertine himself.[23]

In some instances these victims may also pass for audience members, whether from the present day of a given production or from a period within living memory such as Francoist Spain. Rather than tell tales of Others, social outcasts or bewigged aristocrats, such stagings turn a mirror on operagoers themselves, at least those who can afford the elite venues that dominate opera video. Most explicit is Robert Carsen's 2011 version for La Scala, which reproduces the famous interior of the Milan theater onstage. The world of the opera collapses into that of its audience, making way for an exceptionally blunt critique. After Peter Mattei's rough, arrogant, and impeccably attired Don Giovanni descends into smoke (having first been stabbed again, this time by an undead Commendatore), the survivors move downstage to sing the moral, pointing accusing fingers into the house. The video leaves no ambiguity as to their purpose, intercutting shots of the La Scala spectators in their own elegant clothes. But apparently no one has learned anything, for at the end a resurrected Don Giovanni appears and, with a flick of his cigarette, sends the others sinking down in their turn. En-titlement wins again, suggesting that the rich and powerful have no more hope of deliverance than anyone else—and also that they, unlike the down-trodden, bear the responsibility themselves. Upholders rather than victims of modern hierarchies, they help to ensure Don Giovanni's continued exis-tence. Libertinism remains as much a symptom of disorder as a cause.

Family Dysfunction

Much the same holds true when the focus shifts to family, or to inner drama more generally. Don Giovanni's most sinister resurrection may well be the one at the end of the psychologizing Netherlands Opera production dis-cussed in chapter 8.[24] In a setting of bedrooms he is vanquished by beds, symbols of his victims' identities and desires, which physically converge on him as a chandelier falls from above. Freed from his spell and the domestic confines he has dominated, the others then frolic amidst the wreckage—the denouement as pajama party. But he rises again, now dressed like Fran-cisco D'Andrade in Max Slevogt's *Das Champagnerlied*, and rather than flee they greet him gaily, failing to recognize the return of the repressed and its threat to whatever harmony they might reconstitute in its wake. One can imagine a psychoanalytic critic such as Otto Rank muttering a warning: "The many women whom he must always replace anew represent to him the *one* irreplaceable mother; and . . . the rivals and adversaries whom he deceives, defrauds, struggles against, and finally even kills represent the *one* unconquerable moral enemy, the father."[25] As it happens, the Netherlands production actually has an irreplaceable mother, a "Commendatrice" (Peter Lockwood) who sits onstage playing continuo and reading a prayer book. Should all the other symbols fall short, her presence confirms the power of oedipal compulsion, the psychosexual drive that no mere pajama party is going to stop. Animated by unconscious forces, the neurotic Don Giovanni may be even more formidable than his socially privileged counterparts.

Certainly his punishments—the ones that stick, that is—are severe. Without stressing a family theme, Kasper Holten's Royal Opera House production endows Mariusz Kwiecien's Don Giovanni with a rich inner life, revealed by way of live video projections that write, erase, and rewrite the names of his conquests across the set. At the end, though, the names vanish, and with them everything that has constituted his existence. The performance cuts from his final words all the way to the concluding moral, which is piped in from offstage. Alone in the dark, a prisoner of his mind, stripped of human contact and all memory thereof, he loses his very self. Harsher yet is the end of Dmitri Tcherniakov's 2010 production for Aix-en-Provence, where the story plays out among an actual extended family: Donna Elvira is Donna Anna's cousin, Zerlina her daughter, and so on. Bo Skovhus's libertine introduces an experiment in free love, and when it goes awry the others hire an impostor Commendatore to freak him out at a family meeting. It works: after ranging up and down the meeting table, pawing at the other characters, and ripping his shirt open (another version of undressing), he finally sprawls behind his own overturned chair, his spirit broken. One way to defeat a compulsive is to drive him mad.

In this case the musical performance plays an especially prominent role. As led by Louis Langrée, the tempos are fast, the vocal delivery punchy, and the statue's tread a real marching figure. Losing the last of its gravity, the scene rides on an adrenaline rush that carries forward through the più stretto and into the allegro, where the tempos reach 100 bpm and the protagonist finally crashes out. Already depicted as an aging roué and living in a world where time flies (as marked by the intertitles: "one week

---

Worldly Bass Lines

Though the orchestral sound is not so overwhelming as in some other period-instrument recordings, the 2010 video with Louis Langrée and the Freiburg Baroque Orchestra leverages other means to intensify the act 2 finale (*Don Giovanni*, dir. Tcherniakov 2010). Mozart sets several of the statue's incantations over rising bass lines, most strikingly his invitation of Don Giovanni to dinner, in which the voice and the orchestral bass instruments trace a chromatic ascent spanning nearly an octave (DVD disc 2, 1:12:25; chap. 10, 2:14). Played slowly, the ascending scale and the accompanying, fiercely chromatic harmonies deepen the supernatural mystery surrounding the statue's apparition. In a quick alla breve, by contrast, and with the bass foregrounded by balance and mixing, the passage prefigures the heroic mode perfected by some of Mozart's successors, composers such as Luigi Cherubini, Gioachino Rossini, and Ludwig van Beethoven. In Beethoven's *Fidelio*, for example, an even longer chromatic ascent in the bass undergirds the final confrontation between Leonore and Pizzaro. The circumstances remain otherworldly in *Don Giovanni*, but the activated bass makes the real-world aspect of the confrontation, the contest of voices and bodies, feel all the more visceral.

---

later," etc.), Don Giovanni can run no more. In fact, he cannot even leave the stage, remaining on the floor throughout the denouement, frozen in terror as the others taunt and abuse him. When at last they file off, he lies quivering by himself, not dead but undone by the enormity of his rejection. His hell hath no flames, and like the characters of Jean-Paul Sartre's *No Exit* he may wish that his punishment had been more old-fashioned.

At the same time, he might take comfort in how mean-spirited the others look, notably Leporello, whose own expressions of terror are purely sarcastic, and Don Ottavio, who hires the fake Commendatore and drags Donna Elvira back to the table when she attempts to offer comfort. This Don Giovanni is no rogue male, but a member of a family whose gender and generational hierarchies are no less destructive than the inequities of money and rank depicted elsewhere. The moralizing at the end becomes baldly hypocritical, the renunciation of an errant son by relatives who are no better. The finale comes again to depict universal wickedness, negating whatever catharsis its protagonist's fall may once have promised.

And why not? Two and a half centuries after the opera's premiere, "civilized" society has not repudiated the conditions that encourage men to use deception or intimidation, entitlement or violence, in the pursuit of sexual and egoic satisfaction. On the contrary, as so many updated settings labor to show, what used to be aristocratic license lives on in both public and private contexts, buttressed by the thus far inexhaustible resilience of male authority in the face of one critique after another. Even as its hero has become a villain, *Don Giovanni* has still not delivered anyone from injustice, and its more cathartic renderings may simply conceal the fact that this kind of injustice too often goes unchecked—not least by the very people who might look to Mozart's operas for visions of a better world. As James Currie writes about enthusiasm for the composer generally: "So thank you Mozart! But for what exactly? For helping us to preserve the fundamental humanity that glows from out of the heart of our celebration of you, or for helping us to forget the fundamental inhumanity of a world that, in our day-to-day lives, we are inextricably implicated in sustaining?"[26] For *Don Giovanni* to be more than a diversion, a way of avoiding the very issues it depicts, productions may well need to show that everyone in and around it behaves badly, audiences included. Maybe the only route to extinguishing the myth runs through nihilism, a social equivalent of the extreme measures—guns, knives, zombies—now needed to take down the libertine himself.

Or maybe nihilist productions are just exchanging one myth for another. If a Donna Anna who chooses sex becomes a liar, a Don Giovanni who acts no worse than those around him evades responsibility for murder and rape. This is most evident in psychologizing interpretations, where his predations are literally symptoms rather than causes; the man is sick, gripped by urges he can no more resist than the others can avoid the temptation to play along—or, when expectations fall through, to indulge their own worst tendencies. They all channel forces larger than themselves, in which respect some modern Don Giovannis look and sound suspiciously like their Romantic ancestors. Once vessels of eros or biological élan, now they enact the longings of the repressed, the frustrations of the disadvantaged, the corruptions of the powerful. Less admirable than before, they remain all the same not entirely accountable, symbols as much as agents, above or beyond the usual standards of right and wrong. Put another way: for everything that has happened to him in the last few decades, from losing his grandeur to going crazy, Don Giovanni is not yet done with being a myth. Time will tell if he can keep it up.

Acknowledgments

I am grateful to John Gibbs, retired Special Projects Librarian of the University of Washington Music Library, for sparking my initial interest in historical Mozart recordings and facilitating access to the Offenbacher Mozart Collection. I also thank Judy Tsou (Music Librarian Emerita) and Erin Conor (Music Librarian) of the University of Washington Libraries. Marta Tonegutti of the University of Chicago Press has offered wise counsel and encouragement throughout, and Dylan Montanari has been an expert managing editor.

My text has benefited from the insight and generosity of many colleagues. I thank Bonnie Gordon and Mary Hunter for their incisive commentaries on several chapter drafts, and Marcy Linton for advising me on historical costuming. For their thoughtful suggestions and critiques, I am grateful as well to Mary Ann Smart and David Levin, editors of the *Opera Lab* series; to the readers for the University of Chicago Press; and to the respondents to several years' worth of *Don Giovanni* presentations at conferences and colloquia. Special thanks are due to my graduate research assistants at the University of Virginia, Amy Coddington, Craig Comen, and Justin Mueller.

Above all, thank you, Melanie, for your patience and support.

I began research for this book while a fellow at the National Humanities Center in Research Triangle Park, NC, where I was supported by a Frederick Burkhardt Residential Fellowship for Recently Tenured Scholars from the American Council of Learned Societies (2009–10). Further support was provided by a sabbatical year from the University of Virginia (2015–16).

Chapter 8 is a revised version of my article, "Zooming In, Gazing Back: *Don Giovanni* on TV," *Opera Quarterly* 27, no. 1 (2011): 32–65; it is reprinted by permission of Oxford University Press. Portions of chapters 9 and 10 appeared previously in my colloquy contribution, "*Don Giovanni* and the Resilience of Rape Culture," *Journal of the American Musicological Society* 71, no. 1 (2018): 218–22; it is reprinted by permission of the University of California Press.

Notes

Introduction

1. See the discography and videography.

2. On textbooks see Curtis, "Sexual Politics"; Hartford, "Beyond the Trigger Warning," esp. 24–26.

3. The most comprehensive listing is the online *Mozart-Bibliographie* of the Stiftung Mozarteum Salzburg (https://mozarteum.at). Du Mont, *The Mozart–Da Ponte Operas*, 71–128, offers a well-organized guide to literature before 2000.

4. Previous overviews include Bitter, *Wandlungen in den Inszenierungsformen*; Rushton, *W. A. Mozart: "Don Giovanni,"* 66–80; Parakilas, "The Afterlife of *Don Giovanni*"; Gruber, *Mozart and Posterity*, esp. 183–87, 203–7; Andrews, "Don Giovanni: The Transgressor Transgressed."

5. Allanbrook, *Rhythmic Gesture*, 200.

6. Gronda, *Il Don Giovanni*, 77; Stivender and Webb, *"Don Giovanni,"* 373.

7. Tirso de Molina, *The Trickster of Seville*, 188–89; see also the introduction by Gwynne Edwards, xxv–xxvi.

8. See esp. Bitter, *Wandlungen in den Inszenierungsformen*; Rice, *Mozart on the Stage*; Parakilas, "The Afterlife of *Don Giovanni*"; Ellis, "Rewriting *Don Giovanni*"; Cowgill, "Re-gendering the Libertine"; Everist, "Enshrining Mozart"; Woodfield, *Performing Operas for Mozart*, 65–130.

9. Worthen, *Shakespeare Performance Studies*, 7.

10. On the sources see esp. Woodfield, *The Vienna "Don Giovanni"*; Edge, "Mozart's Viennese Copyists," 1742–1866; Mozart, *Don Giovanni*, ed. Plath and Rehm (*Neue Mozart Ausgabe*, 1968, and *Kritischer Bericht*, 2003); Rehm, "Musicological Introduction," in Mozart, *Il dissolute punito ossia Don Giovanni* (facsimile ed., 2009), 3:13–30.

11. Woodfield, *The Vienna "Don Giovanni,"* 1.

12. Ibid., 2; see also Abert, *W. A. Mozart*, 1108–9; Bitter, *Wandlungen in den Inszenierungsformen*, 55–56.

13. For side-by-side synopses see Rushton, *W. A. Mozart: "Don Giovanni,"* 8–26; Woodfield, *The Vienna "Don Giovanni,"* xvi–xvii; Gronda, *Il Don Giovanni*, 119–23.

14. *Dramaturgische Monate*, 1790; quoted and translated in Deutsch, *Mozart: A Documentary Biography*, 354.

15. Littlejohn, *The Ultimate Art*, 120.

16. Rochlitz, *Don Juan*. See also Bitter, *Wandlungen in den Inszenierungsformen*, 87–89; and chaps. 3 and 4 below.

17. Gronda, *Il Don Giovanni*, 8; Don *Giovanni*, dir. Warner 1995; dir. Bieito 2002; dir. Tcherniakov 2010.

18. Ellis, "Rewriting *Don Giovanni*"; on nineteenth-century adaptations see also Bitter, *Wandlungen in den Inszenierungsformen*; Cowgill, "Re-gendering the Libertine"; Possart, *Ueber die Neueinstudierung und Neuszenierung des Mozart'schen Don Giovanni*, 3-10.

19. Rockwell, "A 'Giovanni' Reborn as 'Donna.'"

20. Shaw, "A Non-Mozartian Don," *World*, (May 13, 1891); reprinted in *Shaw's Music*, 2:337.

21. Glass, "*Don Giovanni*—as Perplexing as Ever."

22. Collected in Taruskin, *Text and Act*; see esp. 90–154, 263–97.

23. The most sustained riposte is Butt, *Playing with History*.

24. Cook, *Beyond the Score*, 8–32.

25. Philip, *Performing Music*, 2.

26. Leech-Wilkinson, *The Changing Sound of Music*, chap. 1.2, §18.

27. Philip, *Early Recordings*.

28. On mazurka research, see Cook, *Beyond the Score*, 157–75.

29. Citron, *Opera on Screen*, 2.

30. See also Tambling, *Opera, Ideology and Film*; Schmidt-Sistermanns, *Opernregie im Fernsehen*; Kühnel, "Oper im Fernsehen"; Kühnel, "'Mimesis' und 'Diegesis'"; Tambling, *A Night in at the Opera*; Csobádi, Gruber, and Kühnel, *Das Musiktheater in den audiovisuellen Medien*; Nuy and Lemke, *Theaterbühne, Fernsehbilder*.

31. See esp. Morris, "Digital Diva"; Senici, "Porn Style?"

32. Levin, *Unsettling Opera*, 23–24; Worthen, *Shakespeare Performance Studies*, 3–8.

33. Levin, *Unsettling Opera*, 27. On the intersections of opera scholarship and performance studies see also Smart, "Defrosting Instructions."

34. Poriss, *Changing the Score*; Rutherford, *The Prima Donna*.

35. Parker, *Remaking the Song*, 42–66.

36. Abbate, *In Search of Opera*, xiii.

37. Ibid., 51.

38. Worthen, *Shakespeare Performance Studies*, 11.

39. Abbate, "Music—Drastic or Gnostic?," 510.

40. Ardoin, *The Furtwängler Record*, 106.

41. Cf. *Don Giovanni*, cond. Furtwängler 1954; cond. Furtwängler 1950; cond. Furtwängler 1953. The statue scene is further discussed in chap. 10 below.

42. See Fink, "*Rigoroso* (♪ = 126)."

43. Ross, "The Storm of Style."

44. Cook, *Beyond the Score*, 70 and passim.

45. Brody, "The Revelations of *Don Giovanni*." The production was recorded twice; see *Don Giovanni*, dir. Grandage 2011 and 2016.

46. Littlejohn, *The Ultimate Art*, 125.

47. *Don Giovanni*, dir. Hume 1991; dir. Warner 1995; dir. Zambello 2008.

48. Elizabetta Maschio, in a live recording of 1995 (Kicco KC039, released 1999).

Chapter 1

1. Illustrations in Leppert, *Aesthetic Technologies*, 137–43.

2. *Don Giovanni*, cond. Busch 1936. On the Glyndebourne recording and the challenges of complete opera recording generally in this era, see Gaisberg, *The Music Goes Round*, 165-75.

3. See www.lib.washington.edu/music/specialcollections/mozart.

4. A useful point of comparison is the Mozart Ton- und Filmsammlung of the Stiftung Mozarteum Salzburg (https://mozarteum.at). Its catalogue lists 429 short-playing *Don Giovanni* recordings, and while the proportions are for the most part similar to those found in the Offenbacher collection ("Deh vieni alla finestra" recordings make up 20 percent of the total, those of Don Ottavio, Leporello, and Zerlina, 6–12 percent each), the noblewomen's arias are somewhat better represented (3–6 percent each, as compared to 2–4 percent). This is due in part to more numerous listings for the years 1950–55.

5. Less consistent is the handling of language. Continental European labels typically give titles only in the vernacular, whereas American and British labels tend to give the Italian and an English translation. This presumably reflects stronger Continental traditions of performing opera in translation, both onstage and on recordings.

6. Armstrong, "The Costuming of a Prima Donna"; see also chap. 4 below.

7. Cather, "Three American Singers," 36.

8. *Edison Phonograph Monthly*, January 1911, 21–22.

9. Gronda, *Il Don Giovanni*, 34; Stivender and Webb, "*Don Giovanni*," 327–28; Bishop, *The Libertine*, 22. See also chap. 4 below.

10. Rous, *The Victor Book of the Opera*, 9. Illustrations and further discussion in Leppert, *Aesthetic Technologies*, 97–109.

11. Rous, *The Victor Book of the Opera*, 78–79.

12. E.g., Klein, "The Supremacy of Mozart."

13. *Gramophone*, July 1924, 68.

14. Day, *A Century of Recorded Music*, 76–87.

15. Eisenberg, *The Recording Angel*, 35. See also Katz, *Capturing Sound*, 20–21.

16. E.g., Eisenberg, *The Recording Angel*, 30–34.

17. Mann, *The Magic Mountain*, 646–48.

18. Kittler, *Gramophone, Film, Typewriter*, 51–55.

19. Ibid., 30–38, 87–93.

20. Cf. Katz, *Capturing Sound*, 21–24.

21. *Gramophone*, June 1924, 8.

22. My account of public gramophone listening draws on contemporary reports in *Gramophone* as well as Eisenberg, *The Recording Angel*, 57–70; Lacey, *Listening Publics*.

23. Leppert, *Aesthetic Techologies*, 135.

24. Kenney, *Recorded Music in American Life*, xv. The term "circles of resonance" originated in an essay on dance by Paul Valéry; see ibid., 204 n. 24.

25. Rous, *The Victor Book of the Opera*, 9.

26. Leppert, *Aesthetic Technologies*, 119–22.

27. Turner, "Class, Race, and Uplift in the Opera House."

28. See esp. Miller, *Segregating Sound*.

29. For Anderson's discography, see Keiler, *Marian Anderson*, 353–76.

30. Fleeger, *Mismatched Women*, 1–13.

31. Koestenbaum, *The Queen's Throat*, 49; quoted in Fleeger, *Mismatched Women*, 9.

32. Shuker, *Wax Trash*; see also Eisenberg, *The Recording Angel*, 132.

33. Kenney, *Recorded Music in American Life*, 88–108.

34. See Gelatt, *The Fabulous Phonograph*, 169, 186–87, 248–49.

35. Steane, *The Grand Tradition*, 10.

36. Cuts were the preferred method among instrumentalists. See Katz, *Capturing Sound*, 38–39.

37. See Beardsley, "Speeds and Pitching"; Leech-Wilkinson, *The Changing Sound of Music*, chap. 3.1, §§22–23.

38. Beardsley, "Speeds and Pitching."

39. Trezise, "The Recorded Document," 207.

40. Cf. the discussions in Steane, *The Grand Tradition*, 8–9; and, concerning the much-debated pitch levels of Adelina Patti's recordings, Aspinall, "Adelina Patti," 272–73; Moran, "The Recorded Legacy of Adelina Patti," in Cone, *Adelina Patti*, 313–15; Leech-Wilkinson, *The Changing Sound of Music*, chap. 4, §§11–12; Trezise, "The Recorded Document," 196–97; Jeffrey Miller and Ward Marston, "Note from the Producers," in *The Complete Adelina Patti and Victor Maurel* (Marston Records 52011–2, 1998).

41. The following draws on Leech-Wilkinson, *The Changing Sound of Music*, chap. 3.1, §§3–28; Katz, *Capturing Sound*, 41–44, 56–79; Schmidt Horning, *Chasing Sound*, 11–31.

42. For photographs see Welch and Burt, *From Tinfoil to Stereo*, illustrations 24 and 25.

43. Leppert, *Aesthetic Technologies*, 122–29; Milner, *Perfecting Sound Forever*, 41–49.

Chapter 2

1. Gronda, *Il Don Giovanni*, 52; Stivender and Webb, "*Don Giovanni*," 346.

2. Gronda, *Il Don Giovanni*, 52; Stivender and Webb, "*Don Giovanni*," 346.

3. Ibid.

4. Figs. 2.2 and 2.3 were prepared with the help of Sonic Visualizer, an open-source program available through the Centre for the History and Analysis of Recorded Music (C.H.A.R.M.), http://www.charm.rhul.ac.uk/.

5. On crooning, see McCracken, *Real Men Don't Sing.*

6. See Philip, *Early Recordings*; Philip, *Performing Music in the Age of Recording.*

7. Taruskin, *Text and Act*, 108–11.

8. Cook, *Beyond the Score*, 70, 125.

9. In the recitative, "Chi è là?—Stelle! che vedo!" following the act 1 aria "Ah chi mi dice mai."

10. Kierkegaard, *Either/Or*, 1:301–445.

11. Ibid., 99. See also Williams, "Don Giovanni as an Idea," 32–35; Villadsen, "The Rhetoric of Seduction," 255–57.

12. Taruskin, *Text and Act*, 114.

13. See Cook, *Beyond the Score*, 130–31.

14. Taruskin, *Text and Act*, 114.

15. Hurst, "Collector's Corner," 374.

16. Ibid.

17. Maurel, *Dix ans de carrière*, 347–48.

18. Ibid., 349–53. Maurel is refuting Dumas's preface to Roujon, *Miremonde*, 21–25.

19. On Maurel's stage practice see Henson, "Verdi, Victor Maurel and *fin-de-siècle* Operatic Performance," 64–83.

20. Hurst, "Collector's Corner," 374.

21. Finck, "Maurice Renaud," 614.

22. Hurst, "Collector's Corner," 374.

23. Hurst, "Collector's Corner," 374. I am not aware of a Scotti Serenade recording dating from 1905, but Hurst may have been referring to either the 1903 or the 1909 version released on Victor and Gramophone.

24. *New York Daily Tribune*, December 28, 1899, 6. Many Metropolitan Opera reviews are reprinted or excerpted in the *Met Opera Database*, http://archives .metoperafamily.org/archives/frame.htm.

25. Chuilon, *Mattia Battistini*, 145.

26. *New York Evening Post*, November 30, 1929.

27. *New Republic*, December 18, 1929, 100.

28. Pinza, *An Autobiography*, 132–33.

29. Ibid., 151–54.

30. Walter, *Theme and Variations*, 171–72. On the 1905 Viennese *Don Giovanni* see esp. Baker, "Alfred Roller's Production."

31. *New York Times*, March 21, 1946, 33.

32. Maurel, *Dix ans de carrière*, 353.

33. Roujon, *Miremonde*, 151.

34. Ovid, *The Erotic Poems*, 166–90.

35. *Don Giovanni*, cond. Rudolf 1955.

Chapter 3

1. *Don Giovanni*, dir. Graf 1954.

2. Gronda, *Il Don Giovanni*, 32; Stivender and Webb, "*Don Giovanni*," 326 (amended).

3. Hoffmann, "Don Juan," 25. Further page references in the text.

4. Kierkegaard, *Either/Or*, 1:85.

5. Kalbeck, *Opern-Abende*, 1:31; Heuß, "Das dämonische Element." Similar sentiments were expressed by the soprano Pauline Viardot, a famous Donna Anna and Zerlina in the mid-nineteenth century and the longtime owner of the autograph score of *Don Giovanni*. See Everist, "Enshrining Mozart," 166–67.

6. Oulibicheff, *Nouvelle biographie*, 3:94.

7. Ibid., 3:133.

8. Bitter, *Mozart's "Don Juan*," 94; Bulthaupt, *Dramaturgie der Oper*, 1:182. See also Lert, *Mozart auf dem Theater*, 407.

9. Abert, *W. A. Mozart*, 1049.

10. Ibid., 1050.

11. Ibid., 1073–74.

12. Jahn, *W. A. Mozart*, 4:389.

13. Ibid., 4:390.

14. Busoni, *Von der Einheit der Musik*, 91.

15. Gounod, *Mozart's "Don Giovanni*," 54.

16. Rous, *The Victor Book of the Opera*, 79.

17. Dent, *Mozart's Operas*, 1st ed., 193.

18. Ibid., 257.

19. See Imiela and Roland, *Slevogt und Mozart*; Jung-Kaiser, "'Der weiße d'Andrade.'"

20. Guthmann, *Goldene Frucht*, 265–66.

21. Abert, *W. A. Mozart*, 1074.

22. Walter, *Theme and Variations*, 30; see also 113.

23. Lears, *No Place of Grace*, 13–14. The following draws also on Tosh, *Manliness and Masculinities*, 148–209; Sussman, *Masculine Identities*, 81–98; Beynon, *Masculinities and Culture*, esp. 26–52; Kimmel, *Manhood in America*, 59–136.

24. Hayem, *Le Don Juanisme*, 10.

25. Ibid., 12.

26. Segal, *Slow Motion*, 94–95.

27. Parker, "The Artist's Reward," 31; quoted in Segal, *Slow Motion*, 96.

28. *Adventures of Don Juan*, directed by Vincent Sherman (Warner Bros., 1948). DVD release: Warner Home Video, 2007. *The Affairs of Cellini*, directed by Gregory La Cava (United Artists, 1934). DVD release: 20th Century Fox, 2014.

29. *Carnegie Hall*, directed by Edgar G. Ulmer (United Artists, 1947). DVD release: Bel Canto Society BCS-D0791, 2008.

30. See Schor, *Reading in Detail*.

31. For a summary of more recent views, see Rohringer, "Don Ottavio," 33–39.

32. Dent, *Mozart's Operas*, 1st ed., 238.

33. Abert, *W. A. Mozart*, 1095.

34. Oulibicheff, *Nouvelle biographie*, 3:75–76.

35. Kalbeck, *Mozart's Don Juan*, ix.

36. Gronda, *Il Don Giovanni*, 62; Stivender and Webb, "*Don Giovanni*," 356.

37. Rochlitz, *Don Juan*, 42.

38. Levi, *Mozarts Don Juan*, 70.

39. Grandaur, *Don Juan*, 60–61; cf. Mozart, *Don Giovanni*, ed. Einstein, 460–72.

40. Kalbeck, *Mozart's Don Juan*, 66.

41. A particularly idiosyncratic recording by Beniamino Gigli (b. 1890) seems to show Don Ottavio transitioning to modernism in the course of the aria itself. A sensuous beginning leads to a heroic second verse, after which the first verse returns with noticeably more force. It is as if the middle section awakened some new resolve in the character, a transformation echoed in the powerful final cadences.

42. *Don Giovanni*, dir. Hall 1977; dir. Zambello 2008; dir. Grandage 2011.

43. *Don Giovanni*, dir. Ebert 1961. On Schünemann's editions see Erik Levi, *Mozart and the Nazis*, 77–84. Though they were commissioned for a project of Aryanizing Mozart's Italian operas, the translations frequently retain some or all of Hermann Levi's language: the first verse of "Il mio tesoro intanto," for example, recurs almost verbatim. Cf. Levi, *Mozart and the Nazis*, 83.

44. *Don Giovanni*, dir. Hampe 1987; dir. Strehler 1987.

Chapter 4

1. Maurel, *Dix ans de carrière*, 358.

2. Rochlitz, *Don Juan*, 13.

3. Gronda, *Il Don Giovanni*, 22; Stivender and Webb, "*Don Giovanni*," 315.

4. Rochlitz, *Don Juan*, 14.

5. Gronda, *Il Don Giovanni*, 22; Stivender and Webb, "*Don Giovanni*," 316 (amended).

6. Rochlitz, *Don Juan*, 14.

7. Mozart, *Don Juan: Musique de Mozart*, 61–63. On the origins of the Deschamps/Blaze translation, see Ellis, "Rewriting *Don Giovanni*."

8. Gronda, *Il Don Giovanni*, 34; Stivender and Webb, "*Don Giovanni*," 327–28.

9. Rochlitz, *Don Juan*, 22–23.

10. Gronda, *Il Don Giovanni*, 34; Stivender and Webb, "*Don Giovanni*," 328.

11. Levi, *Mozarts Don Juan*, 38; Rochlitz, *Don Juan*, 23.

12. Oulibicheff, *Nouvelle biographie*, 3:79.

13. Ibid.

14. Ibid., 160–61.

15. Jahn, *W. A. Mozart*, 4:395.

16. Ibid., 4:401.

17. Abert, *W. A. Mozart*, 1061. Otto Rank's psychoanalytical *Die Don-Juan Gestalt* appeared in the same year as Abert's Mozart biography, 1924.

18. Ibid., 1069.

19. Ibid.

20. Regarding "Vedrai carino," Abert writes that Mozart "draws a veil of the most delightful kind over the aria as a whole, preventing the all-too-human from finding expression." Ibid., 1088.

21. Bulthaupt, *Dramaturgie der Oper*, 1:193.

22. Brown-Montesano, *Understanding the Women of Mozart's Operas*, 80.

23. Bulthaupt, *Dramaturgie der Oper*, 1:195.

24. Cohen, *Die dramatische Idee*, 88.

25. Ibid.

26. Allanbrook, *Rhythmic Gesture*, 262. The duettino and "Batti, o bel Masetto" begin in 2/4 and end in 6/8; "Vedrai carino" is in 3/8 but with strong enough two-measure groupings to project a slow 6/8 meter (at least at the right tempo; see below).

27. Columbia GQ 7164, matrix nos. B-4190/B-4212.

28. Aspinall, "Adelina Patti," 270–71; Steane, *The Grand Tradition*, 15.

29. In slower performances of the 2/4, Patti's included, it occasionally sounds as if the governing pulse is the eighth rather than the quarter note, such that the dotted quarters of the 6/8 appear not to accelerate the beat, but to maintain or even slow it down. Nevertheless, the continuity of the cello part through both sections projects a palpable sense of acceleration, its calm accompanimental figures turning into rushing scales and arpeggios.

30. I am grateful to Ian Woodfield for confirming this point (personal communication).

31. Freitas, "Singing Herself," 304.

32. Ibid., 348–52.

33. Ibid., 344–45.

34. Klein, *Herman Klein and the Gramophone*, 46–47.

35. Freitas, "Singing Herself," 352–53.

36. Oulibicheff, *Nouvelle biographie*, 3:134–35.

37. Abert, *W. A. Mozart*, 1069.

38. *Boston Evening Transcript*, April 10, 1908, 13.

39. Freitas, "Singing Herself," 327.

40. *New York Times*, January 6, 1908.

41. See Armstrong, "The Costuming of a Prima Donna."

42. Finck, *Success in Music*, 191.

43. See Finck, "The Evolution of Sex."

44. Finck, *Success in Music*, 183.

45. Finck, "The Evolution of Sex," 3064.

46. Further details of the broadcast are in John Ardoin, liner notes for *Rosa Ponselle on the Air, Vol. 1*, produced by Bill Park and Ward Marston (Marston Records 52012–2, 1999).

47. See esp. the second verse, which begins with laughing challenges and ends with an exaggerated seesawing between fast and slow, loud and soft.

48. Gronda, *Il Don Giovanni*, 57; Stivender and Webb, "*Don Giovanni*," 351.

49. As Allanbrook notes, the passage lends itself to a reading "*alla siciliana*." See *Rhythmic Gesture*, 272–73.

50. On this point, see Allanbrook, *Rhythmic Gesture*, 262–64.

51. As in "Batti batti, o bel Masetto," slower recordings sometimes suggest four medium-tempo beats in the 2/4 (i.e., 4/8), such that the 6/8 appears to increase the pulse by 75 percent on average. The sense of acceleration remains dramatic.

52. Lehmann, *Mein Weg*, 434–35.

53. Ibid., 367.

54. La Grange, *Gustav Mahler*, 289–91.

55. *Don Giovanni*, dir. Himmelmann 2013.

Chapter 5

1. The following overview draws on Schmidt Horning, *Chasing Sound*, 104–39; Brock-Nannestad, "The Development of Recording Technologies," 160–76; Wilentz, *360 Sound*, 127–34, 245–46.

2. On the transition to digital recording, see Milner, *Perfecting Sound Forever*, 104–236.

3. E.g., *Don Giovanni*, cond. Nézet-Séguin 2011.

4. An additional forty-five archival live recordings appear in the listing by Steiger, *Opern-Diskographie*, 296–302.

5. Edited live versions: *Don Giovanni*, cond. Mehta 2009; cond. Nézet-Séguin 2011; cond. Rhorer 2016. Productions released concurrently on video: *Don Giovanni*, cond. Harding 1999; cond. Jacobs 2006.

6. Mozart, *Don Giovanni*, ed. Einstein, XIII–XIV. The B♭s can be heard on *Don Giovanni*, cond. Swarowsky 1950; cond. Giulini 1959; cond. Leinsdorf 1959.

7. For a related account of early Wagner LPs, see Meyer, "Sound Recording," 19–21.

8. *Gramophone*, August 1937, 93.

9. *Gramophone*, December 1959, 304. The recording is Giuseppe Verdi, *Aida*, conducted by Herbert von Karajan (Decca SXL 2167–9).

10. Culshaw, *Ring Resounding*, 5.

11. Adorno, "Opera and the Long-Playing Record," 284–85. See also Meyer, "*Parsifal*'s Aura," 166–68.

12. See *Don Giovanni*, cond. Swarowsky 1950; cond. Moralt 1955; cond. Rudolf 1955; cond. Fricsay 1958.

13. *Don Giovanni*, cond. Barenboim 1974. The title character receives a similarly psychedelic updating in the 1970s reissue of *Don Giovanni*, cond. Fricsay 1958 (Deutsche Grammophon 2728 003). The same company's box set of the following decade (*Don Giovanni*, cond. Karajan 1985) reproduces Slevogt's painting as a photograph, with Samuel Ramey taking the place of D'Andrade.

14. *Don Giovanni*, cond. Maazel 1978. Cf. *Don Giovanni*, cond. Solti 1996; cond. Nézet-Séguin 2011.

15. *Don Giovanni*, cond. Jacobs 2006. Cf. *Don Giovanni*, cond. Barenboim 1991; cond. Kuijken 1995.

16. *Don Giovanni*, cond. Moralt 1955. The booklet offers equally antiquated translations into German (by Hermann Levi, 1897) and French (by Émile Deschamps and Henri Castil-Blaze, 1834).

17. Mozart, *Don Giovanni, An Opera, In Two Acts*, 106. MacFarren's translation had also been used in the G. Schirmer vocal score edition of the opera (New York, 1900), which continues to circulate in a modern reprint (New York: Dover, 2013).

18. Ibid., 21.

19. *Don Giovanni*, cond. Busch 1936, booklet n.p.

20. Murray: *Don Giovanni*, cond. Rudolf 1955; cond. Giulini 1959, cond. Klemperer 1966, cond. Barenboim 1974, cond. Haitink 1984, cond. Muti 1990; Cochrane: *Don Giovanni*, cond. Fricsay 1958, cond. Leinsdorf 1959, cond. Bonynge 1969, cond. Solti 1978; Salter: *Don Giovanni*, cond. Böhm 1967, cond. Davis 1973, cond. Böhm 1977, cond. Karajan 1985, cond. Järvi 1990, cond. Marriner 1990, cond. Gardiner 1994, cond. Abbado 1997; Bardoni: *Don Giovanni*, cond. Östman 1989, cond. Solti 1996, cond. Nézet-Séguin 2011, cond. Currentzis 2015.

21. *Webster's Ninth New Collegiate Dictionary* (Springfield, MA: Merriam-Webster, 1986), 978.

22. Ruppel, "Remarks on Mozart's *Don Giovanni*," in *Don Giovanni*, cond. Böhm 1967, booklet p. 3.

23. Untitled preface, in *Don Giovanni*, cond. Klemperer 1966, booklet n.p.

24. "Essay by Glen Sauls," in *Don Giovanni*, cond. Giulini 1959, booklet p. 3.

25. Hughes, "Mozart's Don Giovanni: 'Scarcely a Human Experience," in *Don Giovanni*, cond. Fricsay 1958, booklet p. 7. Hughes's essay appears in the U.S. release (Deutsche Grammophon DGMO 302); the German release (SLPM 138 050–052) offers reflections by Hans Heinz Stuckenschmidt on the title character as a universal symbol of sensuality.

26. Ibid., 6.

27. Ibid., 5.

28. Kolodin, "A Perspective of *Don Giovanni*," in *Don Giovanni*, cond. Leinsdorf 1959, booklet n.p.

29. Losey and Chris Nelson, "Losey's *Don Giovanni* [interview with Chris Nelson]," in *Don Giovanni*, cond. Maazel 1978, booklet p. 2.

30. Ibid., 3.

31. Higgins, "The Opera of Operas," in *Don Giovanni*, cond. Haitink 1984, booklet p. 13.

32. Ibid., 12.

33. Kunze, "Comedy under the Sign of Catastrophe," in *Don Giovanni*, cond. Harnoncourt 1988, booklet p. 41.

34. Griffiths, "Wolfgang Amadeus Mozart, *Don Giovanni*: Telling and Retelling," in *Don Giovanni*, cond. Östman 1989, booklet pp. 19–20.

35. Ibid., 21.

36. Cf. Meyer, "Sound Recording," 20–21.

37. Landon, "Historical Notes on the Opera," in *Don Giovanni*, cond. Mackerras 1995, booklet pp. 8–12; Landon, "A Work Beyond Its Time," in *Don Giovanni*, cond. Solti 1996, booklet pp. 11–16.

38. Kolodin, "A Perspective of *Don Giovanni*," in *Don Giovanni*, cond. Leinsdorf 1959, booklet n.p.

39. *Don Giovanni*, cond. Furtwängler 1954.

40. *Don Giovanni*, cond. Giulini 1959 (2016 reissue, Warner Classics 0825646994052). *Don Giovanni*, cond. Fricsay 1958 is presented similarly in its 2001 reissue (Deutsche Grammophon 463 629–2), part of a series entitled "The Originals: Legendary Recordings from the Deutsche Grammophon Catalogue."

41. Taruskin, *Text and Act* (see the introduction above).

42. *Don Giovanni*, cond. Jacobs 2006, booklet p. 47.

43. *Don Giovanni*, cond. Barenboim 1974; cond. Kuijken 1995.

44. Culshaw, *Ring Resounding*, 199.

45. *High Fidelity*, November 1958, 45.

46. The system is described in Culshaw, *Ring Resounding*, 12; and Grubb, *Music Makers*, 26. For an illustration see *Don Giovanni*, cond. Klemperer 1966, booklet back cover.

47. Culshaw, *Ring Resounding*, 5.

48. *Don Giovanni*, cond. Leinsdorf 1959.

49. Along similar lines, the Glyndebourne recording of 1936 cranks up a wind machine at the entrance of the statue (*Don Giovanni*, cond. Busch 1936).

50. See esp. *Don Giovanni*, cond. Davis 1973.

51. *Don Giovanni*, cond. Harnoncourt 1988.

52. *High Fidelity*, November 1958, 46.

53. *High Fidelity*, January 1960, 62.

54. Ibid. The recording is *Don Giovanni*, cond. Fricsay 1958.

55. Gronda, *Il Don Giovanni*, 46-47; Stivender and Webb, "*Don Giovanni*," 340–41 (amended).

56. See, for example, the director Jonathan Miller's comments in Romain, *Jonathan Miller*, 63.

57. *Don Giovanni*, cond. Walter 1942.

58. Allen, *Foreign Parts*, 72; for a similar account by the conductor Zubin Mehta, see Romain, *Jonathan Miller*, 207.

59. *Don Giovanni*, cond. Haitink 1984.

60. *Don Giovanni*, cond. Marriner 1990.

61. *Don Giovanni*, cond. Muti 1990. Note especially how Shimmell begins angry and short, "Leporello . . . Vien qui, facciamo pace"; gets suddenly friendly, "prendi"; then returns to business, "quattro doppie."

62. *Don Giovanni*, cond. Currentzis 2015.

63. *Musical Times*, March 1974, 224. The recording is *Don Giovanni*, cond. Davis 1973.

Chapter 6

1. Allanbrook, *Rhythmic Gesture*, 239.

2. Kunze, in *Don Giovanni*, cond. Harnoncourt 1988, booklet p. 41; cf. Kunze, *Mozarts Opern*, 427–30; Dent, *Mozart's Operas*, 1st ed., 245.

3. Lipking, "Donna Abbandonata," 44; Brown-Montesano, *Understanding the Women of Mozart's Operas*, 49. As noted in chap. 5, Paul Griffiths suggests a parallel between Donna Elvira's subjecting herself to humiliation and audiences coming repeatedly back to the opera. See *Don Giovanni*, cond. Östman 1989, booklet p. 21.

4. Dent, *Mozart's Operas*, 1st ed., 244.

5. Hoffmann, *Don Juan*, 19.

6. Adelmann, *Donna Elvira*, 10.

7. Ibid., 27; Maurel, *Dix ans de carrière*, 355; Bulthaupt, *Dramaturgie der Oper*, 1:191–92.

8. Lipking, "Donna Abbandonata," 45; cf. Rushton, *W. A. Mozart: "Don Giovanni,"* 101.

9. Bulthaupt, *Dramaturgie der Oper*, 1:187.

10. Lipking, "Donna Abbandonata," 37. Mariana is the abandoned protagonist of the seventeenth-century *Letters of a Portuguese Nun*. Cf. Brown-Montesano, *Understanding the Women of Mozart's Operas*, 35.

11. Lipking, "Donna Abbandonata," 41; the passage appears in Kierkegaard, *Either/Or*, 1:194.

12. Hotho, *Vorstudien*, 19.

13. Responding to criticisms that "In quali eccessi, o Numi! . . . Mi tradì quell'alma ingrata" impedes the plot, Julian Rushton argues that it fills in Donna Elvira's psychological development between the end of the act 2 sextet, where she discovers Leporello's disguise, and the finale, where she confronts Don Giovanni for the last time. See Rushton, *W. A. Mozart: "Don Giovanni,"* 101–2.

14. Phelan, *Charles Mackerras*, 256.

15. Crutchfield, "The Prosodic Appoggiatura," esp. 270–74.

16. Neumann, "The Appoggiatura," 137; see also Neumann, "Improper Appoggiaturas," 516.

17. *Don Giovanni*, cond. Furtwängler 1950; cond. Furtwängler 1954; cond. Giulini 1959. The 1947 recording is Columbia 72640-D (matrix no. CAX 10054–55).

18. Schwarzkopf, Steane, and Sanders, *Elisabeth Schwarzkopf*, 24.

19. Stivender and Webb, "*Don Giovanni*," 357.

20. Schwarzkopf, Steane, and Sanders, *Elisabeth Schwarzkopf*, 23–24.

21. Helletsgruber: *Don Giovanni*, cond. Busch 1936; Hammond: Columbia DX 1075 (matrix no. CAX 8985–96), 1942; Gadski: Victor 88253 (C 9489), 1910; Lous: Odeon 99046–99047 (xB 3981–82), 1908.

22. In the initial act 2 conversation between Leporello and Don Giovanni, for example, the libertine's self-justification passes without fuss, the continuo remaining steady and John Brownlee declining to shift into arialike lyricism, as so many other Don Giovannis do (see chap. 5 above).

23. *Musical Times* (February 1955): 83 (reviewing HMV's transfer of the recording to LP).

24. Gronda, *Il Don Giovanni*, 49; Stivender and Webb, "*Don Giovanni*," 342.

25. Gronda, *Il Don Giovanni*, 49; Stivender and Webb, "*Don Giovanni*," 343.

26. Gronda, *Il Don Giovanni*, 51–52; Stivender and Webb, "*Don Giovanni*," 345–46.

27. *Gramophone*, March 1937, 434.

28. See esp. Darnton, "Don Juanism from Below"; Dickie, *Cruelty and Laughter*.

29. *Don Giovanni*, cond. Rudolf 1955. Cf. *Don Giovanni*, cond. Krips 1955.

30. *Don Giovanni*, cond. Krips 1955; cond. Fricsay 1958.

31. *Don Giovanni*, cond. Swarowsky 1950.

32. Hughes: *Musical Times* (May 1961): 293; Broder: *High Fidelity*, November 1960, 77.

33. Hughes, ibid. Admittedly, Hughes authored the booklet essay for the competing Deutsche Grammophon recording (*Don Giovanni*, cond. Fricsay 1958).

34. E.g., *Don Giovanni*, cond. Krips 1955; cond. Fricsay 1958; cond. Leinsdorf 1959.

35. See the account in Schwarzkopf, Steane, and Sanders, *Elisabeth Schwarzkopf*, 137.

36. E.g., Schwarzkopf thought the recitatives too hurried; ibid., 23–24.

37. Landon Green, in Gruber, *The Metropolitan Opera Guide*, 304. Cf. the glowing judgment in Blyth, *Opera on CD*, 23–24.

38. *Independent*, July 21, 1995.

39. *Don Giovanni*, cond. Muti 1990; cond. Gardiner 1994; cond. Mackerras 1995.

40. See *Don Giovanni*, dir. Freyer 1998; and https://www.volksoper.at/production/don-giovanni-2015.968939179.en.html (Volksoper Wien, 2015).

Chapter 7

1. Leinsdorf, *The Composer's Advocate*, 101.

2. "*Don Giovanni*: Behind the Scenes," in *Don Giovanni*, dir. Flimm 2001 (at 6:01).

3. Mozart, *Don Giovanni, ossia, Il dissoluto punito*, 1822; Fink, "Ueber das Bedürfniss."

4. See Gerstenberg, "Authentische Tempi" (on Tomašek); Rudolf, "Ein Beitrag zur Geschichte der Temponahme" (on Tomašek and Schlesinger).

5. L. Mozart, *A Treatise*; Türk, *Klavierschule*.

6. On eighteenth-century tempo recommendations see Zaslaw, *Mozart's Symphonies*, 489–501; Brown, *Classical and Romantic*, 282–99.

7. E.g., in Harnoncourt, *The Musical Dialogue*, 92–97; Leinsdorf, *The Composer's Advocate*, 104–5.

8. Harnoncourt, *The Musical Dialogue*, 93–94.

9. Marty, *The Tempo Indications*; Breidenstein, *Mozart's Tempo-System*.

10. Breidenstein, *Mozart's Tempo-System*, 9.

11. Leinsdorf, *The Composer's Advocate*, 107.

12. Ibid., 106–7.

13. Ibid., 104.

14. Harnoncourt, *The Musical Dialogue*, 98–99.

15. Ibid., 98.

16. Leinsdorf: Mozart, *The Marriage of Figaro*, RCA Victor Red Seal LM-6408; Harnoncourt: Mozart, *Le nozze di Figaro*, Teldec 4509–90861–2.

17. E.g., *Don Giovanni*, cond. Böhm 1955 (live); cond. Klemperer 1955 (live); cond. Maazel 1978.

18. E.g., *Don Giovanni*, cond. Rudolf 1955; cond. Fricsay 1958; cond. Muti 1990.

19. Harnoncourt, *The Musical Dialogue*, 100–107.

20. Malloch, "Toward a 'New' (Old) Minuet." Schlesinger also recommended quick tempos for the menuetto, 92 and 100 bpm (table 7.1).

21. Ibid., 16.

22. See esp. the opening figure, whose dotted rhythms are almost inaudible.

23. Gronda, *Il Don Giovanni*, 42; Stivender and Webb, "*Don Giovanni*," 336.

24. Woodfield, *Performing Operas for Mozart*, 101–4.

25. E.g., in *Don Giovanni*, cond. Östman 1989; cond. Norrington 1992; cond. Abbado 1997; cond. de Billy 2002; cond. Currentzis 2015.

26. E.g., *Don Giovanni*, cond. de Billy 2002; cond. Nézet-Séguin 2011; cond. Currentzis 2015.

27. Will, "The Ambivalence of Mozart's Countess."

28. In some cases the acceleration seems to encompass the maestoso before the dance as well, provided its tempo and accentuation project a slowish 2/4 (in which case it sounds slower than the menuetto) and not a moderate 4/8 (in which case it sounds faster).

29. Gronda, *Il Don Giovanni*, 44; Stivender and Webb, "*Don Giovanni*," 338.

30. Ibid.

Chapter 8

1. On reactions to video see Senici, "Porn Style?," 63–67; Morris, "Digital Diva," 96–98; Esse, "Don't Look Now," 81–83. Scholars have also privileged fidelity to the stage; see Catani, *Don Giovanni in Video*, 12–27, 88–99; Kühnel, "Oper im Fernsehen"; Herz, "Von der Bühne auf den Bildschirm."

2. And in some cases via audio mixing; see Senici, "Porn Style?," 67.

3. Clarke, "Like a Virgin?"

4. Wlaschin, *Encyclopedia of Opera on Screen*, 192.

5. On Losey's film see Citron, *Opera on Screen*, 161–204; Kauffmann, "The Abduction from the Theater," 231–34.

6. Retold versions: see *Don Giovanni Unmasked*, dir. Barbara Willis Sweete (2000); *Juan*, dir. Kasper Holten (2010). Mock documentaries: see *Bröderna Mozart*, dir. Suzanne Osten (1986); *The Real Don Giovanni*, dir. Rupert Edwards (1998).

7. For the history of opera on television, see Rose, *Television and the Performing Arts*, 127–88; Citron, *Opera on Screen*, 40–58.

8. Cf. Morris, "Digital Diva," 111–12.

9. Further details are in Schmidt-Sistermanns, *Opernregie im Fernsehen*; Kühnel, "Oper im Fernsehen"; Werckmeister, "Die zweite Regie."

10. Morris, "Digital Diva," 108–10; Sheppard, "Review of the Metropolitan's Opera's New HD Movie Theater Broadcasts"; Heyer, "Live From the Met."

11. Rather than cut at the Commendatore's entrance, Mielke widens the frame suddenly, to similar effect.

12. The directorial responsibility for Sellars's videos is not clear. Werner Vogel is credited as video editor, but Sellars himself took a role in guiding the camera work and presumably in producing the videos. The videos are discussed in Citron, *Opera on Screen*, 205–48; Pfeiffer, "'Directed for Video'"; Elste, "Vom Mythos der Realität."

13. For further details on the editing of all four videos, see Will, "Zooming In, Gazing Back," 36–40, esp. tables 2 and 4.

14. *Don Giovanni*, dir. Zeffirelli 2000; dir. Pasqual 2005.

15. Kunze, *Mozarts Opern*, 388.

16. Quoted above, p. 4.

17. Miller, *Don Giovanni: Myths*, ix.

18. On the classic format of live television see Auslander, *Liveness*, 11–24; Ellis, *Visible Fictions*, 132–37; Caughie, *Television Drama*, 122.

19. Abert, *W. A. Mozart*, 1057.

20. Maurel, *Dix ans de carrière*, 358.

21. Rank, *The Don Juan Legend*, 125.

22. Brophy, *Mozart the Dramatist*.

23. *Amadeus*, dir. Milos Forman, screenplay by Peter Shaffer (Warner Bros. Pictures, 1984).

24. Miller, *Don Giovanni: Myths*, viii.

25. Allanbrook, *Rhythmic Gesture*, 223.

26. Ibid., 207–8; cf. Kerman, "Reading *Don Giovanni*," 117–25.

27. Caughie, *Television Drama*, 161.

28. Wiesmann and Müller, "Gespräch mit Brian Large," 49.

29. Citron, *Opera on Screen*, 224.

30. Whether it is appropriate for viewers to see vocal mechanics is a perennial subject for debate; see Kühnel, "Oper im Fernsehen," 184; Esse, "Don't Look Now," 81–82.

31. A notable variation appears in *Don Giovanni*, dir. Marelli 1988 (filmed by Hugo Käch), in which several arias feature close-ups fading in over full-stage views of the singers, until one sees both images at once. The slowly materializing close-ups have a dreaminess reminiscent of dissolves, giving an unusually strong impression that windows are being opened on the characters' interiority.

32. Berlioz, *Memoirs*, 65.

33. Gronda, *Il Don Giovanni*, 69; Stivender and Webb, "*Don Giovanni*," 364.

34. See Citron, "The Elusive Voice"; Citron, "Subjectivity in the Opera Films of Ponnelle."

35. Gronda, *Il Don Giovanni*, 5; Stivender and Webb, "*Don Giovanni*," 297.

36. *Don Giovanni*, dir. Zeffirelli 1958.

37. *Don Giovanni*, dir. Vaccari 1960.

38. *Don Giovanni*, dir. Zeffirelli 1990; dir. Himmelmann 2013.

39. The classic account is Mulvey, "Visual Pleasure."

40. See Blier, "The Fit Lady Sings"; Tommasini, "Should the Fat Lady Diet"; Tommasini, "Second Date With a Little Black Dress."

41. While giving first place to voice, Wayne Koestenbaum's reflections on fandom also attest to the importance of singers' images, notably in photographs. See *The Queen's Throat*, 26–29, 51–52, passim.

42. Lapsley and Westlake, *Film Theory*, 225–27; Klinger, "In Retrospect: Film Studies Today."

43. Morley, *Television*, 1–58.

44. Altman, "Television/Sound."

45. Ellis, *Visible Fictions*, 138.

46. For Jeremy Tambling, this emphasis on liveness encourages another order of fetishizing, of the operatic institution and the privilege it represents; *A Night In At the Opera*, 6–9.

47. Abbate, "Opera; or, The Envoicing of Women," 235.

48. Belton, "The Bionic Eye"; Schmidt-Sistermanns, *Opernregie im Fernsehen*, 104.

Chapter 9

1. Hoffmann, *Don Juan*, 27–28. Further page references given parenthetically in the text.

2. Lert, *Mozart auf dem Theater*, 399; Mann, *The Operas of Mozart*, 468.

3. Fanny Lewald, *National-Zeitung*, Nr. 81 (1860); quoted in Wolzogen, *Wilhelmine Schröder-Devrient*, 163–65.

4. Abert, *W. A. Mozart*, 1061; cf. Oulibicheff, *Nouvelle biographie*, 3:91–92; Jahn, *W. A. Mozart*, 4:416 n.; Wolzogen, *Wilhelmine Schröder-Devrient*, 165–72.

5. Abert, *W. A. Mozart*, 1061.

6. Rank, *The Don Juan Legend*, 92, 101.

7. Jouve, *Le Don Juan de Mozart*, 70.

8. Ibid., 105–6.

9. Einstein, *Mozart*, 439; Patrick Cairns "Spike" Hughes, in *Don Giovanni*, cond. Fricsay 1958, booklet p. 7.

10. Rosenberg, *Don Giovanni*, 270–72; Barraud, *Les cinq grands opéras*, 24–26, 70.

11. Moberly, *Three Mozart Operas*, 151.

12. Ibid., 148, 160.

13. Mann, *The Operas of Mozart*, 468.

14. Dent, *Mozart's Operas*, 1st ed., 235.

15. Dent, *Mozart's Operas*, 2nd ed., 158.

16. Mann, *The Operas of Mozart*, 468.

17. An exception is the commentary by Hughes, in *Don Giovanni*, cond. Fricsay 1958; see above, pp. 116–17.

18. Rushton, *W. A. Mozart: "Don Giovanni,"* 60; Osborne, *The Complete Operas of Mozart*, 268.

19. Clément, *Opera*, 34.

20. Ibid., 35.

21. Kristeva, *Tales of Love*, 193.

22. Ford, *Così?*, 185. Ford would later strengthen his critique; see his *Music, Sexuality and the Enlightenment*, 104–5. On Kunze and Griffiths see chap. 5 above.

23. Brown-Montesano, *Understanding the Women of Mozart's Operas*, 2; see also 25–33.

24. Curtis, "Sexual Politics," 128; see also 125.

25. For the evolution of textbook accounts, compare the excerpts quoted in Curtis, "Sexual Politics," 140–42, with Forney, Dell'Antonio, and Machlis, *The Enjoyment of Music*, 188–89, or Kelly, *Music Then and Now*, 176–207.

26. Shea, *Acting in Opera* (1915), and Challis, "The Technique of Operatic Acting" (1927), illustrate and defend the "pose"-based style still very much evident in mid-century videos.

27. *Don Giovanni*, dir. Meyer 1960; dir. Vaccari 1960. Though the acting is more dynamic, *Don Giovanni*, dir. Felsenstein 1966, featuring Klara Barlow, leaves a similar impression of confidence and strength.

28. *Don Giovanni*, dir. Ebert 1961.

29. Similar stagings are found in *Don Giovanni*, dir. Graf 1954 (also with Grümmer, now singing in Italian); dir. Zeffirelli 1958 (with Ilva Ligabue); dir. Browning 1960 (with Leontyne Price).

30. While a history of this change remains to be written, its impact may be measured by comparing Shea, *Acting in Opera*, and Challis, "The Technique of Operatic Acting," to more recent how-to manuals such as Goldovsky, *Bringing Opera to Life* (1968); Helfgot, *The Third Line* (1993); and Hicks, *Singer and Actor* (2011).

31. *Don Giovanni*, dir. Strehler 1987.

32. *Don Giovanni*, dir. Mariani 1997.

33. Hunter, *Mozart's Operas*, 216–19.

34. Ibid., 223. The productions are *Don Giovanni*, dir. Sellars 1990; dir. Bieito 2002.

35. *Don Giovanni*, dir. Warner 1995.

36. Andrews, "Don Giovanni: The Transgressor Transgressed," 123.

37. Ibid., 122.

38. *Don Giovanni*, dir. Zeffirelli 1958.

39. *Don Giovanni*, dir. Reinhardt 2011.

40. Ferguson, "Rape and the Rise of the Novel," 88–98, explores the central role of victim testimony in rape cases.

41. Higgins, *The Making of an Opera*, 83–86.

42. *Don Giovanni*, cond. Maazel 1978, booklet p. 2.

43. *Don Giovanni*, dir. Zeffirelli 1990; dir. Hampe 1987. Hampe argues that the character is driven by guilt; see his *Alles Theater*, 75.

44. Gronda, *Il Don Giovanni*, 29; my translation.

45. "Making of: Talking Don Giovanni," in *Don Giovanni*, dir. Kušej 2006 (at 5:22).

46. *Don Giovanni*, dir. Bechtolf 2006.

47. *Don Giovanni*, dir. Brook 2002.

48. *Don Giovanni*, dir. Tcherniakov 2010.

49. *Don Giovanni*, dir. Bechtolf 2014.

50. "Interview: Peter Brook," in *Don Giovanni*, dir. Brook 2002 (at 6:13, 12:30).

51. "Commentary with Director Kasper Holten and Set Designer Es Devlin," in *Don Giovanni*, dir. Holten 2014 (DVD 1, at 1:05:16).

52. *Don Giovanni*, dir. Guth 2008.

53. "Commentary with Director Kasper Holten and Set Designer Es Devlin," in *Don Giovanni*, dir. Holten 2014 (DVD 1, at 11:06).

54. Sutcliffe, *Believing in Opera*, 131–32.

55. Ibid., 131.

56. Wright, "Consuming Passions," 322.

57. Mila, *Lettura del Don Giovanni*, 68.

Chapter 10

1. Gronda, *Il Don Giovanni*, 81; Stivender and Webb, "*Don Giovanni*," 376 (amended).

2. On the production of the film, including the contributions of Furtwängler's conducting assistants, see Schönhaar, "Mozart als Filmkomponist?," 410–15; Ardoin, *The Furtwängler Record*, 107. The film is further discussed by Ruth E. Müller in Catani, *Don Giovanni in Video*, 88–99.

3. In his review of the cinema release, Arthur Jacobs complained of poor sound quality and inconsistent volume levels; *Sight and Sound* 25, no. 1 (Summer 1955): 35. While the video remastering solves many issues, the sound differs markedly between the recitatives and the concerted music, and there are occasional synchronization problems.

4. *Gramophone*, May 1986 (https://www.gramophone.co.uk/review/mozart-don -giovanni-44). The recording is *Don Giovanni*, cond. Furtwängler 1954.

5. Bulthaupt, *Dramaturgie der Oper*, 1:200–201.

6. *Don Giovanni*, dir. Hampe 1987.

7. Higgins, *The Making of an Opera*, 67.

8. For related examples, see *Don Giovanni*, dir. Marelli 1988; dir. Bondy 1990; dir. Hume 1991; dir. de Simone 1999.

9. *Don Giovanni*, dir. Browning 1960.

10. Allen, *Foreign Parts*, 54.

11. *Don Giovanni*, dir. Schaaf 1992. On Schaaf's conception see Allen, *Foreign Parts*, 174–77.

12. *Don Giovanni*, dir. Strehler 1987.

13. *Destination Mozart: A Night at the Opera with Peter Sellars*, directed by Andrea Simon, at 25:33 (Arcadia Pictures, 1990; DVD release: Kultur D1288, n.d.).

14. Both the stage and the video versions of Sellars's Mozart trilogy (*The Marriage of Figaro*, *Don Giovanni*, and *Così fan tutte*) have inspired considerable commentary; see esp. Littlejohn, *The Ultimate Art*, 130–55; Allanbrook, "Zerlina's 'Batti, batti'"; Citron, *Opera on Screen*, 205–48; Pfeiffer, "'Directed for Video'"; Elste, "Vom Mythos der Realität."

15. Parakilas, "The Afterlife of *Don Giovanni*," 264.

16. The opening scene is purposely vague, as is the synopsis provided by Sellars in the video edition: "the girl's father (the 'Commendatore'), waking to the sound of screams, challenges the intruder on his doorstep. Fifty seconds later he is dead." Booklet accompanying *Don Giovanni*, dir. Sellars 1990, 8.

17. André, "From Otello to Porgy," 25.

18. *Don Giovanni*, dir. Guth 2008; dir. Kent 2010; dir. Himmelmann 2013.

19. *Don Giovanni*, dir. Reinhardt 2011.

20. *Don Giovanni*, dir. Bechtolf 2014. Cf. *Don Giovanni*, dir. Carsen 2011, led by Daniel Barenboim in a style reminiscent of his studio recordings of 1975 and 1991.

21. Kierkegaard, *Either/Or*, 1:119; see also Williams, "Don Giovanni as an Idea," 35–36.

22. "*Don Giovanni* Unmasked," in *Don Giovanni*, dir. Kent 2010 (at 10:10).

23. *Don Giovanni*, dir. Pasqual 2005.

24. *Don Giovanni*, dir. Wieler and Morabito 2007.

25. Rank, *The Don Juan Legend*, 41.

26. Currie, "Postmodern Mozart," 215.

Discography

Part 1. Excerpt Audio Recordings

Part 1 of this Discography lists the recordings discussed in chapters 1–4. For a complete catalog of the recordings I consulted, see the Offenbacher Mozart Collection of the University of Washington Library: https://www.lib.washington.edu/music/specialcollections/mozart; http://db.lib.washington.edu/offenbacher/. Labels and catalog numbers refer to the discs held by the Offenbacher Mozart Collection. The recorded performances themselves, identified by matrix number, were in many cases released or reissued on multiple labels. For information on original or alternate releases, as well as dates of recording and accompanying performers, see the collection catalogue and the discographies cited in the Bibliography (incl. *C.H.A.R.M. Discography*; *Discography of American Historical Recordings*; Bauer, *The New Catalogue*; Hogarth, "A Critical Survey"; Lustig, "Charles Gilibert"; Thoma, "Maria Cebotari"). The language of performance is Italian unless otherwise noted.

Abbreviations: (P) piano, (C) conductor, *mx* matrix number

BATTI BATTI, O BEL MASETTO

1904	Marcella Sembrich. Victor 85038, *mx* C-1954.
1905	Adelina Patti. Landon Ronald (P). HMV 03055, *mx* 541f.
1906	Marcella Sembrich. Victor 88026, *mx* C-3359.
1907	Luisa Tetrazzini. Percy Pitt (C). Gramophone 053148, *mx* 2180f.
1908	Geraldine Farrar. Victor Orchestra. Victor 88126, *mx* C-6064.

1911	Luisa Tetrazzini. HMV DB 537, *mx* ac 5168f.
1913	Selma Kurz. Opera Disc 72645, *mx* 684m.
1915	Lola Artôt de Padilla. Grammophon 65217, *mx* 915m. German.
1917	Marie Sundelius. Explanatory talk by Harry E. Humphrey. Edison 82120, *mx* 4868-C-1–3 (aria), 5459-C-3–3 (explanatory talk).
1917	Elisabeth Schumann. Odeon 76726, *mx* xxB 6380. German.
1920	Elisabeth Schumann. Polydor 65655, *mx* 276as. German.
1926	Elisabeth Schumann. George Byng (C). HMV DB 946, *mx* CC8404.
1928	Ninon Vallin. Pathé X 7167, *mx* 20442. French.
1929	Irene Eisinger. Ultraphon A139, *mx* 10107. German.
1931	Adele Kern. Decca PO 5059, *mx* 2877 BH. German.
1934	Rosa Ponselle. André Kostelanetz (C). WABC Broadcast.
1937	Lucrezia Bori. Frank Black (C). Victor 14614. *mx* CS-011414.
1941	Maria Cebotari. Gerhard Steeger (C). Orchester des Deutschen Opernhauses Berlin. Deutsche Grammophon 67686 LM, *mx* 1342 1/2. German.
1942	Bidu Sayão. Erich Leinsdorf (C). Columbia 71577-D, *mx* XCO 35291.
n.d.	Anna Maria Guglielmetti. Columbia GQ 7164, *mx* B-4190.

DEH VIENI ALLA FINESTRA

1901	Jean Lassalle. IRCC 3034, *mx* A-10326. French.
1902	Mattia Battistini. HMV VA 5, *mx* 442z.
1902	Jean-François Delmas. Gramophone 2–32639, *mx* 1290-F RP. French.
1902	Antonio Scotti. Gramophone 2–2707, *mx* 2020 nb.
1903	Leopold Demuth. Gramophone 2–42732, *mx* 889z. German.
1903	Antonio Scotti. Columbia 1207, *mx* 1207.
1903	Antonio Scotti. Gramophone 052113, *mx* 875.
1904	Mario Ancona. Gramophone 52130, *mx* 2320L.
1904	Victor Maurel. IRCC 9, *mx* XPh 66.
1904	Mario Sammarco. Fonotipia 39001, *mx* XPh 7.
1905	Mario Ancona. Pathé 4309, *mx* 4309.
1906	John Forsell. Gramophone 2–82410, *mx* 2706e. Swedish.
1907	Charles Gilibert. Columbia 30096, *mx* 30096 (M 129–1). French.
1907	Charles Gilibert. Columbia 30121, *mx* 30121 (M130–1). French.
1907	Giuseppe De Luca. Fonotipia 39940, *mx* XPh 2312.
1907	Titta Ruffo. Gramophone 2–52625, *mx* 10914b.
1908	Maurice Renaud. HMV D 851, *mx* 877i. French.
1909	Antonio Scotti. Victor Orchestra. Victor 88194, *mx* C-8281.

1909   Mario Sammarco. Percy Pitt (C). Gramophone 2–052013, *mx* 3743f.

1913   Mattia Battistini. Carlo Sabajno (C). Gramophone 2–52847, *mx* 17569b.

1920   Titta Ruffo. Victor 87112, *mx* B-12620.

1922   Heinrich Rehkemper. Gramophone 62402, *mx* 1954ar. German.

1924   Mattia Battistini. HMV AGSA 30, *mx* Bk 1433.

1925   Heinrich Schlusnus. Gramophone 70704, *mx* 2236ax. German.

1926   Heinrich Schlusnus. Gramophone 62560, *mx* 508bk. German.

1926   Mariano Stabile. Columbia C12528, *mx* B1136.

1927   Vanni-Marcoux (Jean-Émile Diogène). Piero Coppola (C). Gramophone DA 937, *mx* BTR 3190.

1928   Gerhard Hüsch. Frieder Weissmann (C), Staatsoper Berlin. Decca G-20248, *mx* P 36458–o. German.

1928   Luigi Montesanto. Columbia 12288, *mx* B2104.

1929   Roger Bourdin. Gustave Cloëz (C). Odeon 188663, *mx* KI 2442. French.

1930   Ezio Pinza. Victor 1467, *mx* BVE-59733.

1934   Willi Domgraf-Fassbaender. HMV EG 2906, *mx* OD 1785. German.

1934   André Pernet. Piero Coppola (C). Gramophone DA 4850, *mx* OPG 1451–1. French.

1935   Heinrich Schlusnus. Leo Blech (C), Orchester der Staatsoper Berlin. Polydor 62760, *mx* GR 6055. German.

1937   Karl Schmitt-Walter. Orchester der Staatsoper. Telefunken A 2223, *mx* 21978. German.

1942   Tito Gobbi. Umberto Berrettoni (C). HMV DA 5430, *mx* OBA 5149.

1947   Paul Schöffler. Decca M 613, *mx* DR 11578.

1953   Giuseppe Valdengo. Cetra AT 0368, *mx* 55510 FC.

1955   Giuseppe Taddei. Cetra AT 0403, *mx* 11636.

n.d.   Sándor (Alexander) Svéd. MHV MK 1593, *mx* M 1437. Hungarian.

FIN CH'HAN DAL VINO

1902   Mattia Battistini. HMV AGSA 24, *mx* 439z.

1902   Antonio Scotti. Gramophone 2–2707, *mx* 2020 nb.

1903   Antonio Scotti. Columbia 1207, *mx* 1207.

1904   Antonio Scotti. Victor 85031, *mx* C-1094.

1906–7   Francisco D'Andrade. Parlophone PO 69, *mx* 2–280.

1909   Leopold Demuth. Gramophone GC 4–42230, *mx* 14647u. German.

1914   Titta Ruffo. Victrola 938, *mx* B14274.

1926   Heinrich Schlusnus. Gramophone 62560, *mx* 630bk. German.

1926   Mariano Stabile. Columbia C12528, *mx* B1138.

1928   Gerhard Hüsch. Frieder Weissmann (C), Staatsoper Berlin. Decca G 20248, *mx* P 36459–0. German.

1930   Ezio Pinza. Victor 1467, *mx* BVE-59732.

1937   Karl Schmitt-Walter. Orchester der Staatsoper. Telefunken A 2223, *mx* 21975. German.

IL MIO TESORO INTANTO

1907   Giuseppe Anselmi. Fonotipia 62167, *mx* XPh 2571.

1908   Hermann Jadlowker. Odeon Rxx 76024, *mx* xxB 4057. German.

1908   Fernando De Lucia. HMV DA 124, *mx* 11166b.

1916   Hermann Jadlowker. Gramophone 042511, *mx* 966m. German.

1916   John McCormack. Victor 74484, *mx* C-17647.

1922   Charles Hackett. Columbia 98047, *mx* 98047.

1929   Charles Hackett. Columbia 7218-M, *mx* 98663.

1930   Koloman von Pataky. Polydor 95374, *mx* 214 BV.

1935   Peter Anders. Hans Schmidt-Isserstedt (C), Orchester der Reichsoper Berlin. Telefunken E 1796, *mx* 20569. German.

1935   Julius Patzak. Deutsche Grammophon 35029, *mx* 547 1/2 GS. German.

1939   Beniamino Gigli. HMV DB 3809, *mx* 2EA 7909.

1939   Richard Tauber. Walter Goehr (C). Decca 16001, *mx* WXE 9806 T2A.

1945   Charles Kullman. Columbia Test Pressing, *mx* xco 35021.

1946   Max Lichtegg. Gideon Fagan (C). London Test Pressing, *mx* SAR-391.

1948   Petre Munteanu. Mario Rossi (C), Orchestra sinfonica di Torino. Cetra CB 20503, *mx* 2–71205.

LÀ CI DAREM LA MANO

1902   Ellen Brandt-Forster and Leopold Demuth. Gramophone 44180, *mx* 1086x. German.

1905   Irene de Bohuss and Mario Sammarco. Fonotipia 39298, *mx* XPh 351.

1906   Emilia Corsi and Mattia Battistini. Carlo Sabajno (C). Gramophone 054104, *mx* 882c.

1907   Elise Elizza and Friedrich Weidemann. Gramophone GC 2–44165, *mx* 10051u. German.

1907   Eva Tetrazzini and Luigi Baldassare. Phonodisc Mondial 97.

1908   Geraldine Farrar and Antonio Scotti. Victor 89015, *mx* C-5014.

1911   Lydia Lipkowska and Ramón Blanchart. Columbia 30718, *mx* 30718.

1920   Lotte Lehmann and Heinrich Schlusnus. Polydor 72933, *mx* 155 ap. German.

1932   Emmy Bettendorf and Gerhard Hüsch. Frieder Weissmann (C). Decca 20010, *mx* 133563. German.

1935   Margherita Perras and Gerhard Hüsch. Victor 4374, *mx* ORA 398. German.

1936   Erna Berger and Heinrich Schlusnus. Clemens Krauss (C), Orchester der Staatsoper Berlin. Polydor 62760, *mx* 2933 1/2. German.

1939   Elisabeth Rethberg and Ezio Pinza. HMV DA 1950, *mx* OA 036864.

1941   Toti Dal Monte and Augusto Beuf. HMV DA 5406, *mx* OBA 4376.

1943   Gwen Catley and Dennis Noble. Warwick Braithwaite (C), Hallé Orchestra. HMV B 9338, *mx* OER 693. English.

1952   Elfride Trötschel and Horst Günter. Ferdinand Leitner (C), Württemburgisches Staatsorchester. Deutsche Grammophon DG 36087, *mx* 3886 STN. German.

VEDRAI CARINO

1915   Lola Artôt de Padilla. Gramophone 65217, *mx* 914m. German.

1931   Adele Kern. Decca PO 5059, *mx* 2876 BH. German.

1932   Yvonne Brothier. Edouard Bervily (C). Gramophone DB 4878, *mx* 2PG 166–2. French.

1947   Nadine Conner. Max Rudolf (C), Metropolitan Opera Orchestra. Columbia 73072-D, *mx* xco 39415.

1941   Maria Cebotari. Orchester des Deutschen Opernhauses Berlin. Deutsche Grammophon 67686 LM, *mx* 1343 1/2. German.

1942   Bidu Sayão. Erich Leinsdorf (C). Columbia 71577-D, *mx* XCO 32592.

1952   Elisabeth Schwarzkopf. John Pritchard (C), Philharmonia Orchestra. Columbia LB 145, *mx* CA 22824–5D.

1955   Erika Köth. Wilhelm Schüchter (C). Electrola DA 5525, *mx* ORA 7684–1. German.

Part 2. Complete Audio Recordings

The date of the recording appears first, with the date of release following the label and catalog number. For studio recordings and studio-produced live recordings, reissues are listed where I have been unable to consult original releases. For additional archival live recordings, as well as further information on original and alternate releases, see Steiger, *Opern-Diskographie*, 296–302.

Abbreviations: *DG* Don Giovanni, *DA* Donna Anna, *DE* Donna Elvira, *DO* Don Ottavio, *L* Leporello, *Z* Zerlina, *M* Masetto, *C* Commendatore, *P* producer(s)

STUDIO RECORDINGS AND STUDIO-PRODUCED LIVE RECORDINGS

1936   Fritz Busch, Glyndebourne Festival Orchestra, Glyndebourne Festival Chorus. HMV DB 2961–2983, 1936. Reissues: RCA Victor LCT 6102, n.d. (LP); EMI CHS 7 61030 2, 1988 (CD). *DG* John Brownlee, *DA* Ina Souez, *DE* Luise Helletsgruber, *DO* Koloman von Pataky, *L* Salvatore Baccaloni, *Z* Audrey Mildmay, *M* Roy Henderson, *C* David Franklin, *P* David Bricknell.

1950   Hans Swarowsky, Vienna Symphony Orchestra, Chorus of the Vienna State Opera. Haydn Society HSLP 2030, 1951. *DG* Mariano Stabile, *DA* Gertrude Grob-Prandl, *DE* Hilde Konetzni, *DO* Herbert Handt, *L* Alois Pernerstorfer, *Z* Hedda Heusser, *M* Alfred Poell, *C* Oskar von Czerwenka, *P* Richard Wadleigh.

1955   Josef Krips, Vienna Philharmonic Orchestra, Chorus of the Vienna State Opera. Decca LXT 5103–5106, 1956. *DG* Cesare Siepi, *DA* Suzanne Danco, *DE* Lisa Della Casa, *DO* Anton Dermota, *L* Fernando Corena, *Z* Hilde Gueden, *M* Walter Berry, *C* Kurt Böhme, *P* Victor Olof, Peter Andry.

1955   Rudolf Moralt, Wiener Symphoniker, Wiener Kammerchor. Philips A 00280–00282 L, 1955. Reissue: Preiser Records MONO 90762, 2009. *DG* George London, *DA* Hilde Zadek, *DE* Sena Jurinac, *DO* Léopold Simoneau, *L* Walter Berry, *Z* Graziella Sciutti, *M* Eberhard Wächter, *C* Ludwig Weber, *P* Marius van der Meulen.

1955   Max Rudolf, Orchestra Sinfonica di Torino della RAI. Coro di Torino della RAI. Cetra C 1253, 1956. Reissue: Warner Fonit 0927 43561, 2002. *DG* Giuseppe Taddei, *DA* Maria Curtis Verna, *DE* Carla Gavazzi, *DO* Cesare Valletti, *L* Italo Tajo, *Z* Elda Ribetti, *M* Vito Susca, *C* Antonio Zerbini.

1956   Hans Rosbaud, Orchestre de la Société des Concerts du Conservatoire, Chœur du Festival d'Aix en Provence. Pathé DTX 218–221, 1957.

Reissue: Great Opera Performances GOP 66.356, 2006. *DG* Antonio Campo, *DA* Teresa Stich-Randall, *DE* Suzanne Danco, *DO* Nicolai Gedda, *L* Marcello Cortis, *Z* Anna Moffo, *M* André Vessières, *C* Raffaele Ariè.

1958   Ferenc Fricsay, Radio-Symphonie-Orchester Berlin, RIAS Kammerchor. Deutsche Grammophon SLPM 138 050–52, 1959. Reissue: Deutsche Grammophon 463 629–2, 2001. *DG* Dietrich Fischer-Dieskau, *DA* Sena Jurinac, *DE* Maria Stader, *DO* Ernst Haefliger, *L* Karl Christian Kohn, *Z* Irmgard Seefried, *M* Ivan Sardi, *C* Walter Kreppel, *P* Otto Gerdes.

1959   Carlo Maria Giulini, Philharmonia Orchestra, Philharmonia Chorus. EMI SLS 5083, 1960. *DG* Eberhard Wächter, *DA* Joan Sutherland, *DE* Elisabeth Schwarzkopf, *DO* Luigi Alva, *L* Giuseppe Taddei, *Z* Graziella Sciutti, *M* Piero Cappuccilli, *C* Gottlob Frick, *P* Walter Legge.

1959   Erich Leinsdorf, Wiener Philharmoniker, Wiener Staatsopernchor. RCA Victor Red Seal LSC-6410, 1960. *DG* Cesare Siepi, *DA* Birgit Nilsson, *DE* Leontyne Price, *DO* Cesare Valletti, *L* Fernando Corena, *Z* Eugenia Ratti, *M* Heinz Blankenburg, *C* Arnold van Mill, *P* Eric Smith.

1966   Otto Klemperer, New Philharmonia Orchestra, New Philharmonia Chorus. Angel SDL-3700, 1966. *DG* Nicolai Ghiaurov, *DA* Claire Watson, *DE* Christa Ludwig, *DO* Nicolai Gedda, *L* Walter Berry, *Z* Mirella Freni, *M* Paolo Montarsolo, *C* Franz Crass, *P* Peter Andry, Suvi Raj Grubb.

1967   Karl Böhm, Orchester des Nationaltheaters Prag, Tschechischer Sängerchor Prag. Deutsche Grammophon 2711 006, 1967. *DG* Dietrich Fischer-Dieskau, *DA* Birgit Nilsson, *DE* Martina Arroyo, *DO* Peter Schreier, *L* Ezio Flagello, *Z* Reri Grist, *M* Alfredo Mariotti, *C* Martti Talvela, *P* Hans Hirsch.

1969   Richard Bonynge, English Chamber Orchestra, Ambrosian Singers. London OSA 1434, 1970. *DG* Gabriel Bacquier, *DA* Joan Sutherland, DE Pilar Lorengar, *DO* Werner Krenn, *L* Donald Gramm, *Z* Marilyn Horne, *M* Leonardo Monreale, *C* Clifford Grant.

1973   Colin Davis, Orchestra of the Royal Opera House Covent Garden, Chorus of the Royal Opera House Covent Garden. Philips 6707 022/6998 008, 1973. *DG* Ingvar Wixell, *DA* Martina Arroyo, *DE* Kiri Te Kanawa, *DO* Stuart Burrows, *L* Wladimiro Ganzarolli, *Z* Mirella Freni, *M* Richard Van Allan, *C* Luigi Roni.

1974   Daniel Barenboim, English Chamber Orchestra, Scottish Opera Chorus. Angel SDL 3811, 1974. *DG* Roger Soyer, *DA* Antigone Sgourda, *DE* Heather Harper, *DO* Luigi Alva, *L* Geraint Evans, *Z* Helen Donath, *M* Alberto Rinaldi, *C* Peter Lagger, *P* Suvi Raj Grubb.

1977   Karl Böhm, Wiener Philharmoniker, Wiener Staatsopernchor. Deutsche Grammophon 2740 194/2709 085, 1978. *DG* Sherrill Milnes, *DA* Anna Tomowa-Sintow, *DE* Teresa Zylis-Gara, *DO* Peter Schreier, *L* Walter Berry, *Z* Edith Mathis, *M* Dale Düsing, *C* John Macurdy, *P* Werner Mayer.

1978   Lorin Maazel, Orchestra of the Théâtre National de l'Opéra Paris, Chorus of the Théâtre National de l'Opéra Paris. CBS Masterworks M3 35192, 1979. *DG* Ruggero Raimondi, *DA* Edda Moser, *DE* Kiri Te Kanawa, *DO* Kenneth Riegel, *L* José van Dam, *Z* Teresa Berganza, *M* Malcolm King, *C* John Macurdy, *P* Paul Myers.

1978   Georg Solti, London Philharmonic Orchestra, London Opera Chorus. London OSA-1444, 1979. *DG* Bernd Weikl, *DA* Margaret Price, *DE* Sylvia Sass, *DO* Stuart Burrows, *L* Gabriel Bacquier, *Z* Lucia Popp, *M* Alfred Sramek, *C* Kurt Moll, *P* Christopher Raeburn.

1981   Libor Pešek, Prague Chamber Orchestra, Prague National Theatre Chorus. Supraphon 1116 3531–3, 1982. Reissue: Supraphon 3296–2 612, 1997. *DG* Václav Zitek, *DA* Eva Děpoltová, *DE* Magdaléna Hajóssyová, *DO* Vojtech Kocián, *L* Karel Berman, *Z* Jana Jonášová, *M* Dalibor Jedlička, *C* Eduard Haken, *P* Jana Smékalová.

1984   Bernard Haitink, London Philharmonic Orchestra, Glyndebourne Chorus. EMI 7 47037 8, 1984. *DG* Thomas Allen, *DA* Carol Vaness, *DE* Maria Ewing, *DO* Keith Lewis, *L* Richard Van Allan, *Z* Elizabeth Gale, *M* John Rawnsley, *C* Dimitri Kavrakos, *P* John Fraser.

1985   Herbert von Karajan, Berliner Philharmoniker, Chor der Deutschen Oper Berlin. Deutsche Grammophon 419 179–2, 1986. *DG* Samuel Ramey, *DA* Anna Tomowa-Sintow, *DE* Agnes Baltsa, *DO* Gösta Winbergh, *L* Ferruccio Furlanetto, *Z* Kathleen Battle, *M* Alexander Malta, *C* Paata Burchuladze, *P* Günther Breest.

1985   Rafael Kubelik, Symphonie-Orchester des Bayerischen Rundfunks, Chor des Bayerischen Rundfunks. Eurodisc 302 435–445, 1985. *DG* Alan Titus, *DA* Julia Varady, *DE* Arleen Auger, *DO* Thomas Moser, *L* Rolando Panerai, *Z* Edith Mathis, *M* Rainer Scholze, *C* Jan-Hendrik Rootering, *P* Theodor Holzinger, Friedrich Welz.

1988   Nikolaus Harnoncourt, Royal Concertgebouw Orchestra Amsterdam, Chor der Niederländischen Oper. Teldec 2292–44184–2, 1989. *DG* Thomas Hampson, *DA* Edita Gruberová, *DE* Roberta Alexander,

DO Hans Peter Blochwitz, *L* László Polgár, *Z* Barbara Bonney, *M* Anton Scharinger, *C* Robert Holl, *P* Wolfgang Mohr, Helmut Mühle.

1989 Arnold Östman, Drottningholm Court Theatre Orchestra, Drottningholm Court Theatre Chorus. L'Oiseau-Lyre 425 943–2, 1990. *DG* Håkan Hagegård, *DA* Arleen Auger, *DE* Della Jones, *DO* Nico van der Meel, *L* Gilles Cachemaille, *Z* Barbara Bonney, *M* Bryn Terfel, *C* Kristinn Sigmundsson, *P* Peter Wadland, Michael Haas.

1990 Neeme Järvi, Kölner Rundfunk-Sinfonie-Orchester, Kölner Rundfunkchor. Chandos 8920–22, 1992. *DG* Renato Bruson, *DA* Sona Ghazarian, *DE* Gertrud Ottenthal, *DO* Giuseppe Sabbatini, *L* Nicola Ghiuselev, *Z* Patrizia Pace, *M* Stefano Rinaldi-Miliani, *C* Franco de Grandis, *P* Michael Horwath, Heiner Müller-Adolphi.

1990 Neville Marriner, Academy of St. Martin in the Fields, Ambrosian Opera Chorus. Philips 432 129–2, 1991. *DG* Thomas Allen, *DA* Sharon Sweet, *DE* Karita Mattila, *DO* Francisco Araiza, *L* Simone Alaimo, *Z* Marie McLaughlin, *M* Claudio Otelli, *C* Robert Lloyd, *P* Erik Smith.

1990 Riccardo Muti, Wiener Philharmoniker, Konzertvereinigung Wiener Staatsopernchor. EMI 7 54255 2, 1991. *DG* William Shimell, *DA* Cheryl Studer, *DE* Carol Vaness, *DO* Frank Lopardo, *L* Samuel Ramey, *Z* Susanne Mentzer, *M* Natale de Carolis, *C* Jan-Hendrik Rootering, *P* David Groves.

1991 Daniel Barenboim, Berliner Philharmoniker, RIAS Kammerchor. Erato 2292–45588–2, 1992. Reissue: Warner Classics & Jazz, 2564 67700–7, 2011. *DG* Ferruccio Furlanetto, *DA* Lella Cuberli, *DE* Waltraud Meier, *DO* Uwe Heilmann, *L* John Tomlinson, *Z* Joan Rodgers, *M* Michele Pertusi, *C* Matti Salminen, *P* Helmut Mühle.

1992 Roger Norrington, London Classical Players, Schütz Choir of London. EMI 7–54859–2, 1993. Reissue: Virgin Classics 7243 5 61601 2 5, 1999. *DG* Andreas Schmidt, *DA* Amanda Halgrimson, *DE* Lynne Dawson, *DO* John Mark Ainsley, *L* Gregory Yurisich, *Z* Nancy Argenta, *M* Gerald Finley, *C* Alastair Miles, *P* David R. Murray.

1994 John Eliot Gardiner, English Baroque Soloists, Monteverdi Choir. Deutsche Grammophon Archiv 445 870–2, 1995. *DG* Rodney Gilfry, *DA* Luba Orgonasova, *DE* Charlotte Margiono, *DO* Christoph Prégardien, *L* Ildebrando d'Arcangelo, *Z* Eirian James, *M* Julian Clarkson, *C* Andrea Silvestrelli, *P* Peter Czornyj, Karl-August Naegler.

1995 Sigiswald Kuijken, La Petite Bande, Collegium Compostellanum. Accent 95116–8, 1996. Reissue: Brilliant Classics 92877, 2006. *DG* Werner van Mechelen, *DA* Elena Vink, *DE* Christina Högman,

DO Markus Schäfer, *L* Huub Claessens, *Z* Nancy Argenta, *M* Nancy de Vries, *C* Harry van der Kamp, *P* Adelheid Glatt, Andreas Glatt.

1995   Charles Mackerras, Scottish Chamber Orchestra, Scottish Chamber Chorus. Telarc CD-80420, 1996. Reissue: Telarc CD-80726, 2008. *DG* Bo Skovhus, *DA* Christine Brewer, *DE* Felicity Lott, *DO* Jerry Hadley, *L* Alessandro Corbelli, *Z* Nuccia Focile, *M* Umberto Chiummo, *C* Umberto Chiummo, *P* James Mallinson.

1996   Jean-Claude Malgoire, La Grande Écurie et la Chambre du Roy. Astrée E8635, 1998. *DG* Nicolas Rivenq, *DA* Danielle Borst, *DE* Véronique Gens, *DO* Simon Edwards, *L* Huub Claessens, *Z* Sophie Marin Degor, *M* Patrick Donnelly, *C* Patrick Donnelly, *P* Dominique Daigremont.

1996   Georg Solti, London Philharmonic Orchestra, London Voices. London 455 500–2, 1997. *DG* Bryn Terfel, *DA* Renée Fleming, *DE* Ann Murray, *DO* Herbert Lippert, *L* Michele Pertusi, *Z* Monica Groop, *M* Roberto Scaltriti, *C* Mario Luperi, *P* Michael Woolcock.

1997   Claudio Abbado, Chamber Orchestra of Europe, Coro di Ferrara Musica. Deutsche Grammophon 457 601–2, 1998. *DG* Simon Keenlyside, *DA* Carmela Remigio, *DE* Soile Isokoski, *DO* Uwe Heilmann, *L* Bryn Terfel, *Z* Patrizia Pace, *M* Ildebrando d'Arcangelo, *C* Matti Salminen, *P* Christopher Alder, John Fisher.

1998   Alain Lombard, Orchestra della Svizzera Italiana, Coro della Radio Svizzera. Forlane 316787, 1999. *DG* Boris Martinovich, *DA* Patrizia Pace, *DE* Michèle Lagrange, *DO* Domenico Ghegghi, *L* Marcos Fink, *Z* Liliana Nikiteanu, *M* Davide Baronchelli, *C* Anatoli Kotscherga, *P* Ivan Pastor.

1999   Daniel Harding, Mahler Chamber Orchestra, Soloistes de l'Académie européenne de musique d'Aix-en-Provence. Virgin Classics 7243 5 45425 2 7, 2000. Reissue: Virgin Classics 50999 9 55536 2 7, 2012. *DG* Peter Mattei, *DA* Carmela Remigio, *DE* Véronique Gens, *DO* Mark Padmore, *L* Gilles Cachemaille, *Z* Lisa Larsson, *M* Till Fechner, *C* Gudjon Oskarsson, *P* Alain Lanceron, Daniel Zalay.

2000   Michael Halász, Nicolaus Estherházy Sinfonia, Hungarian Radio Chorus. Naxos 8.660080–82, 2001. *DG* Bo Skovhus, *DA* Adrianne Pieczonka, *DE* Regina Schörg, *DO* Torsten Kerl, *L* Renato Girolami, *Z* Ildikó Raimondi, *M* Boaz Daniel, *C* Janusz Monarcha, *P* Ibolya Tóth.

2002   Bertrand de Billy, Vienna Radio Symphony Orchestra, Chorus Sine Nomine. Arte Nova Classics 82876 86721 2, 2006. *DG* Kwangchul

Youn, *DA* Regina Schörg, *DE* Heidi Brunner, *DO* Jeffrey Francis, *L* Maurizio Muraro, *Z* Birgid Steinberger, *M* Reinhard Mayr, *C* Reinhard Hagen, *P* Andrea Seebohm, Michael Lewin, Malgorzata Kragora.

2006   René Jacobs, Freiburger Barockorchester, RIAS Kammerchor. Harmonia Mundi 901664.66, 2007. *DG* Johannes Weisser, *DA* Olga Pasichnyk, *DE* Alexandrina Pendatchanska, *DO* Kenneth Tarver, *L* Lorenzo Regazzo, *Z* Sunhae Im, *M* Nikolay Borchev, *C* Alessandro Guerzoni, *P* Richard Lorber.

2009   Zubin Mehta, Israel Philharmonic Orchestra, New Israeli Vocal Ensemble. Helicon Classics 02–9627, 2009. *DG* Nicola Ulivieri, *DA* Anna Samuil, *DE* Maria Luigia Borsi, *DO* Dmitry Korchak, *L* Maurizio Muraro, *Z* Chen Reiss, *M* Simon Orfila, *C* Marco Spotti, *P* Yaron Karshai.

2011   Yannick Nézet-Séguin, Mahler Chamber Orchestra, Vocalensemble Rastatt. Deutsche Grammophon 477 9878, 2012. *DG* Ildebrando d'Arcangelo, *DA* Diana Damrau, *DE* Joyce DiDonato, *DO* Rolando Villazón, *L* Luca Pisaroni, *Z* Mojca Erdmann, *M* Konstantin Wolff, *C* Vitalij Kowaljow, *P* Ute Fesquet, Daniel Zalay.

2015   Teodor Currentzis, Musicaeterna (Orchestra and Chorus of the Perm Opera and Ballet Theatre). Sony Classics 88985316032, 2016. *DG* Dimitris Tiliakos, *DA* Myrtò Papatanasiu, *DE* Karina Gauvin, *DO* Kenneth Tarver, *L* Vito Priante, *Z* Christina Gansch, *M* Guido Loconsolo, *C* Mika Kares, *P* Bogdan Roščić, Tessa Fanelsa.

2016   Jérémie Rhorer, Le Cercle de l'harmonie, Chœur de Radio France. Alpha Classics 379, 2017. *DG* Jean-Sébastian Bou, *DA* Myrtò Papatanasiu, *DE* Julie Boulianne, *DO* Julien Behr, *L* Robert Gleadow, *Z* Anna Grevelius, *M* Marc Scoffoni, *C* Steven Humes, *P* Elsa Biston.

ARCHIVAL LIVE RECORDINGS

1934   Tullio Serafin, Metropolitan Opera Orchestra, Metropolitan Opera Chorus. Andromeda ANDRCD 9026, 2008. *DG* Ezio Pinza, *DA* Rosa Ponselle, *DE* Maria Müller, *DO* Tito Schipa, *L* Virgilio Lazzari, *Z* Editha Fleischer, *M* Louis D'Angelo, *C* Emanuel List.

1942   Bruno Walter, Metropolitan Opera Orchestra, Metropolitan Opera Chorus. Andromeda ANDRCD 9040, 2008. *DG* Ezio Pinza, *DA* Rose Bampton, *DE* Jarmila Novotná, *DO* Charles Kullman, *L* Alexander Kipnis, *Z* Bidu Sayão, *M* Mack Harell, *C* Norman Cordon.

1943   Paul Breisach, Metropolitan Opera Orchestra, Metropolitan Opera

Chorus. Guild Historical GHCD 2236–7, 2003. *DG* Ezio Pinza, *DA* Zinka Milanov, *DE* Jarmila Novotná, *DO* James Melton, *L* Salvatore Baccaloni, *Z* Bidu Sayão, *M* Mack Harell, *C* Norman Cordon, *P* Jonathan Wearn.

1943   Karl Elmendorff, Dresden State Opera Orchestra, Dresden State Opera Chorus. Preiser Records/Paperback Opera MONO 20023, 2003. *DG* Mathieu Ahlersmeyer, *DA* Marianne Schech, *DE* Margarete Teschemacher, *DO* Hans Hopf, *L* Kurt Böhme, *Z* Elfriede Weidlich, *M* Gottlob Frick, *C* Heinrich Pflanzl.

1944   Georg Szell, Metropolitan Opera Orchestra, Metropolitan Opera Chorus. Golden Melodram GM 5.0043, 2002. *DG* Ezio Pinza, *DA* Florence Kirk, *DE* Eleanor Steber, *DO* Charles Kullman, *L* Salvatore Baccaloni, *Z* Bidu Sayão, *M* Mack Harell, *C* Norman Cordon.

1948   Kurt Schröder, Orchester des Frankfurter Rundfunks, Chor des Frankfurter Rundfunks. Cantus Classics CACD 5.00829, 2005. *DG* Karl Schmitt-Walter, *DA* Clara Ebers, *DE* Felice Hüni-Mihacsek, *DO* Franz Fehringer, *L* Otto von Rohr, *Z* Maria Madlen-Madsen, *M* Xaver Waibel, *C* Helmut Fehn.

1950   Wilhelm Furtwängler, Wiener Philharmoniker, Chor der Wiener Staatsoper. Opera d'Oro OPD 1298, 2000. *DG* Tito Gobbi, *DA* Ljuba Welitsch, *DE* Elisabeth Schwarzkopf, *DO* Anton Dermota, *L* Erich Kunz, *Z* Irmgard Seefried, *M* Alfred Poell, *C* Josef Greindl.

1950   Hans Rosbaud, Orchestra of the Aix-en-Provence Festival, Chorus of the Aix-en-Provence Festival. Pathé (Connaisseur) GM 6.0016, 2004. *DG* Renato Capecchi, *DA* Carla Castellani, *DE* Suzanne Danco, *DO* Léopold Simoneau, *L* Marcello Cortis, *Z* Emmy Loose, *M* Eraldo Coda, *C* Raffaele Ariè.

1952   Hans Rosbaud, Orchestre de la Société des Concerts du Conservatoire, Ensemble Vocale de Paris. Walhall Eternity Series WLCD 0168, 2006. *DG* Heinz Rehfuss, *DA* Carla Martinis, *DE* Leonie Rysanek, *DO* Léopold Simoneau, *L* Marcello Cortis, *Z* Pierette Alarie, *M* Eraldo Coda, *C* Raffaele Ariè.

1953   Wilhelm Furtwängler, Salzburg Festival Orchestra, Salzburg Festival Chorus. Music & Arts Program of America CD 1129, 2004. *DG* Cesare Siepi, *DA* Elisabeth Grümmer, *DE* Elisabeth Schwarzkopf, *DO* Anton Dermota, *L* Otto Edelmann, *Z* Erna Berger, *M* Walter Berry, *C* Raffaele Ariè.

1954   Wilhelm Furtwängler, Wiener Philharmoniker, Chor der Wiener Staatsoper. EMI 3 36799 2, 1986. *DG* Cesare Siepi, DA Elisabeth

Grümmer, *DE* Elisabeth Schwarzkopf, *DO* Anton Dermota, *L* Otto
Edelmann, *Z* Erna Berger, *M* Walter Berry, *C* Deszö Ernster, *P* Victor
Olof, Peter Andry.

1954   Max Rudolf, Metropolitan Opera Orchestra, Metropolitan Opera
Chorus. Walhall Eternity Series WLCD 0237, 2008. *DG* George Lon-
don, *DA* Margaret Harshaw, *DE* Eleanor Steber, *DO* Eugene Conley,
L Fernando Corena, *Z* Nadine Conner, *M* Lorenzo Alvary, *C* Luben
Vichegonov.

1955   Karl Böhm, Orchestra of the Teatro di San Carlo di Napoli, Chorus
of the Teatro di San Carlo di Napoli. Connaisseur GM 6.0010, 2001.
DG Giuseppe Valdengo, *DA* Birgit Nilsson, *DE* Sena Jurinac, *DO*
Anton Dermota, *L* Sesto Bruscantini, *Z* Alda Noni, *M* Walter Berry,
C Gottlob Frick.

1955   Otto Klemperer, Cologne Radio Symphony Orchestra, Cologne Radio
Symphony Chorus. Testament SBT 2149, 1999. *DG* George London,
DA Hilde Zadek, *DE* Maud Cunitz, *DO* Léopold Simoneau, *L* Benno
Kusche, *Z* Rita Streich, *M* Horst Günter, *C* Ludwig Weber.

1956   Dimitri Mitropoulos, Wiener Philharmoniker, Chor der Wiener
Staatsoper. Sony SM3K 64263, 1994. *DG* Cesare Siepi, *DA* Elisabeth
Grümmer, *DE* Lisa Della Casa, *DO* Léopold Simoneau, *L* Fernando
Corena, *Z* Rita Streich, *M* Walter Berry, *C* Gottlob Frick, *P* Hans
Sachs.

1957   Karl Böhm, Metropolitan Opera Orchestra, Metropolitan Opera Cho-
rus. Andromeda ANDRCD 9026, 2008. *DG* Cesare Siepi, *DA* Eleanor
Steber, *DE* Lisa Della Casa, *DO* Jan Peerce, *L* Fernando Corena,
Z Roberta Peters, *M* Theodor Uppman, *C* Giorgio Tozzi.

1959   Karl Böhm, Metropolitan Opera Orchestra, Metropolitan Opera Cho-
rus. Walhall Eternity Series/Archipel WLCD 0275, 2009. *DG* George
London, *DA* Eleanor Steber, *DE* Lisa Della Casa, *DO* Cesare Valletti,
L Ezio Flagello, *Z* Laurel Hurley, *M* Theodor Uppman, *C* William
Wilderman.

1960   Michael Gielen, Orchestra of the Teatro Sao Carlos, Chorus of
the Teatro Sao Carlos. Connaisseur GM 6.0013, 2001. *DG* Eberhard Wächter, *DA* Teresa Stich-Randall, *DE* Montserrat Caballé,
DO Waldemar Kmentt, *L* Erich Kunz, *Z* Lisa Otto, *M* Albrecht Peter,
C Hans Hoffmann.

1960   John Pritchard, Royal Philharmonic Orchestra London, Glyndebourne
Festival Chorus. Golden Melodram GM 5.0077, 2011. *DG* Ernest
Blanc, *DA* Joan Sutherland, *DE* Ilva Ligabue, *DO* Richard Lewis,

L Sesto Bruscantini, *Z* Mirella Freni, *M* Leonardo Monreale, *C* Marco Stefanoni.

1960   Wolfgang Sawallisch, Gürzenich-Orchester Köln, Chor der Oper Köln. Deutsche Grammophon/Universal 476 3676, 2009. *DG* Hermann Prey, *DA* Elisabeth Grümmer, *DE* Hildegard Hillebrecht, *DO* Fritz Wunderlich, *L* Georg Stern, *Z* Edith Mathis, *M* Hans-Georg Knoblich, *C* Franz Crass, *P* Kleopatra Sofroniou.

1961   Ferenc Fricsay, Orchestra of the German Opera, Chorus of the German Opera. Golden Melodram GM 5.0059, 2006. *DG* Dietrich Fischer-Dieskau, *DA* Elisabeth Grümmer, *DE* Pilar Lorengar, *DO* Donald Grobe, *L* Walter Berry, *Z* Erika Köth, *M* Ivan Sardi, *C* Josef Greindl.

1962   Joseph Keilberth, Münchner Philharmoniker der Bayerischen Staatsoper, Münchner Chor der Bayerischen Staatsoper. Golden Melodram GM 5.0041, 2003. *DG* George London, *DA* Hildegard Hillebrecht, *DE* Sena Jurinac, *DO* Nicolai Gedda, *L* Benno Kusche, *Z* Anneliese Rothenberger, *M* Albrecht Peter, *C* Gottlob Frick.

1962   Georg Solti, Covent Garden Orchestra, Covent Garden Opera Chorus. Medici Arts/BBC ROHS 007, 2007. *DG* Cesare Siepi, *DA* Leyla Gencer, *DE* Sena Jurinac, *DO* Richard Lewis, *L* Geraint Evans, *Z* Mirella Freni, *M* Robert Savoie, *C* David Ward, *P* John Pattrick, Edward Blakeman.

1963   Herbert von Karajan, Wiener Philharmoniker, Chor der Wiener Staatsoper. Gala GL 100.608, 1998. *DG* Eberhard Wächter, *DA* Leontyne Price, *DE* Hilde Güden, *DO* Fritz Wunderlich, *L* Walter Berry, *Z* Graziella Sciutti, *M* Rolando Panerai, *C* Walter Kreppel.

1967   Karl Böhm, Metropolitan Opera Orchestra, Metropolitan Opera Chorus. BellaVoce BLV 107.403, 1999. *DG* Cesare Siepi, *DA* Joan Sutherland, *DE* Pilar Lorengar, *DO* Nicolai Gedda, *L* Ezio Flagello, *Z* Laurel Hurley, *M* Theodor Uppman, *C* Bonaldo Giaotti.

1970   Carlo Maria Giulini, RAI Symphony Orchestra, Chorus of Rome. Opera d'Oro OPD 1144, 1998. *DG* Nicolai Ghiaurov, *DA* Gundula Janowitz, *DE* Sena Jurinac, *DO* Alfredo Kraus, *L* Sesto Bruscantini, *Z* Olivera Miljakovic, *M* Walter Monachesi, *C* Dimitri Petkov.

1970   Herbert von Karajan, Wiener Philharmoniker, Konzertvereinigung Wiener Staatsopernchor. Orfeo C 615 033 D, 2003. *DG* Nicolai Ghiaurov, *DA* Gundula Janowitz, *DE* Teresa Zylis-Gara, *DO* Stuart Burrows, *L* Geraint Evans, *Z* Olivera Miljakovic, *M* Rolando Panerai, *C* Victor von Halem.

1973   Wolfgang Sawallisch, Orchester der Bayerischen Staatsoper, Chor der Bayerischen Staatsoper. Myto MCD 053.313, 2005. *DG* Ruggero Raimondi, *DA* Margaret Price, *DE* Julia Varady, *DO* Hermann Winkler, *L* Stafford Dean, *Z* Lucia Popp, *M* Enrico Fissore, *C* Kurt Moll.

1973   Wolfgang Sawallisch, Orchester der Bayerischen Staatsoper, Chor der Bayerischen Staatsoper. Myto MCD 053.313, 2005. *DG* Ruggero Raimondi, *DA* Margaret Price, *DE* Julia Varady, *DO* Hermann Winkler, *L* Stafford Dean, *Z* Lucia Popp, *M* Enrico Fissore, *C* Kurt Moll.

∗

Videography

The date of the recording appears first. If applicable, the date of release follows the label and catalog number. The language of performance is Italian unless otherwise noted.

Abbreviations: (D) stage director, (F) film/television/video director, (C) conductor, *DG* Don Giovanni, *DA* Donna Anna, *DE* Donna Elvira, *DO* Don Ottavio, *L* Leporello, *Z* Zerlina, *M* Masetto, *C* Commendatore

1954 Herbert Graf (D), Paul Czinner (F), Wilhelm Furtwängler (C), Wiener Philharmoniker, Chor der Wiener Staatsoper. Salzburg Festival. Deutsche Grammophon 0730199, 1994/2001. *DG* Cesare Siepi, *DA* Elisabeth Grümmer, *DE* Lisa Della Casa, *DO* Anton Dermota, *L* Otto Edelmann, *Z* Erna Berger, *M* Walter Berry, *C* Deszö Ernster.

1958 Franco Zeffirelli (D), Nino Sanzogno (C), Orchestra Alessandro Scarlatti di Napoli della Rai, Coro del Teatro di San Carlo. Teatro San Carlo, Naples. Unreleased broadcast. *DG* Mario Petri, *DA* Orietta Moscucci, *DE* Ilva Ligabue, *DO* Luigi Alva, *L* Sesto Bruscantini, *Z* Graziella Scuitti, *M* Franco Calabrese, *C* Ferruccio Mazzoli.

1960 Kirk Browning (D), Peter Herman Adler (C), NBC Opera Orchestra. NBC Opera Theater, New York. Unreleased broadcast. *DG* Cesare Siepi, *DA* Leontyne Price, *DE* Helen George, *DO* Charles K. L. Davis, *L* James Pease, *Z* Judith Raskin, *M* John Reardon, *C* John Macurdy.

1960 Jean Meyer (D), Alberto Erede (C), Orchestre de la Société des

Concerts du Conservatoire. Aix-en-Provence Festival. Unreleased broadcast. *DG* Gabriel Bacquier, *DA* Teresa Stich-Randall, *DE* Shakeh Vartenissian, *DO* Luigi Alva, *L* Rolando Panerai, *Z* Mariella Adani, *M* Ugo Trama, *C* Giorgio Tadeo.

1960 Giacomo Vaccari (D), Francesco Molinari-Pradelli (C), Orchestra Sinfonica della RAI di Milano, Coro della RAI di Milano. Radiotelevisione Italiana, Milan. Video Artists International DVD4314, 2005. *DG* Mario Petri, *DA* Teresa Stich-Randall, *DE* Leyla Gencer, *DO* Luigi Alva, *L* Sesto Bruscantini, *Z* Graziella Sciutti, *M* Renato Cesari, *C* Heinz Borst.

1961 Carl Ebert (D), Rolf Unkel (F), Ferenc Fricsay (C), Orchester der Deutsche Oper Berlin, Chor der Deutsche Oper Berlin. Deutsche Oper, Berlin. Arthaus Musik 101574, 2011. German. *DG* Dietrich Fischer-Dieskau, *DA* Elisabeth Grümmer, *DE* Pilar Lorengar, *DO* Donald Grobe, *L* Walter Berry, *Z* Erika Köth, *M* Ivan Sardi, *C* Josef Greindl.

1966 Walter Felsenstein (D), Georg Mielke (F), Zdeněk Kosler (C), Orchester der Komische Oper Berlin, Chor der Komische Oper Berlin. Komische Oper, Berlin. Arthaus Musik 101299, 2009. German. *DG* György Melis, *DA* Klara Barlow, *DE* Anny Schlemm, *DO* John Moulson, *L* Rudolf Asmus, *Z* Eva-Maria Baum, *M* Fritz Hübner, *C* Herbert Rössler.

1977 Heinz Arnold (D), Otmar Suitner (C), Orchester der Staatsoper Berlin, Chor der Staatsoper Berlin. NHK Hall, Tokyo. Unreleased broadcast. *DG* Theo Adam, *DA* Anna Tomowa-Sintow, *DE* Kay Griffel, *DO* Peter Schreier, *L* Siegfried Vogel, *Z* Renate Hoff, *M* Peter Olesch, *C* Motomu Itsuki.

1977 Peter Hall (D), Dave Heather (F), Bernard Haitink (C), London Philharmonic Orchestra, Glyndebourne Festival Chorus. Glyndebourne Festival. Arthaus Musik 101087, 2005. *DG* Benjamin Luxon, *DA* Horiana Branisteanu, *DE* Rachel Yakar, *DO* Leo Goeke, *L* Stafford Dean, *Z* Elizabeth Gale, *M* John Rawnsley, *C* Pierre Thau.

1978 Herbert Graf (D), Kirk Browning (F), Richard Bonynge (C), Metropolitan Opera Orchestra. Metropolitan Opera House, New York. Unreleased broadcast. *DG* James Morris, *DA* Joan Sutherland, *DE* Julia Varady, *DO* John Brecknock, *L* Gabriel Bacquier, *Z* Huguette Tourangeau, *M* Allan Monk, *C* John Macurdy.

1980 Vittorio Patané (D), Jose Luis Mendizabal (F), Alexander Sander (C), Orquestra Simfònica del Gran Teatre del Liceu, Cor del Gran Teatre del Liceu. Gran Teatro del Liceu, Barcelona. Unreleased broadcast. *DG* Nelson Portella, *DA* Luisa Bosabalian, *DE* Adelaide Negri, *DO* Raimundo Mettre, *L* Pedro Liendo, *Z* Carla Basto, *M* Vicenç Esteve, *C* Antonio Borras.

1981   Jean-Pierre Vincent (D) and Jean Dautremay (D), John Pritchard (C), Scottish Chamber Orchestra, Chœurs du Festival Aix-en-Provence. Aix-en-Provence Festival. Lyric Distribution Video. Unreleased broadcast. *DG* Michael Devlin, *DA* Ellen Shade, *DE* Isobel Buchanan, *DO* Barry McCauley, *L* Malcolm King, *Z* Christine Porta, *M* Paolo Martinelli, *C* Dimitri Kavrakos.

1984   Karl-Ernst Herrmann (D), Gérard Mortier (F), Sylvain Cambreling (C), Symfonie-Orkest en Koor van de Nationale Opera. La Monnaie, Brussels. Unreleased broadcast. *DG* José van Dam, *DA* Ashley Putnam, *DE* Christiane Eda-Pierre, *DO* Stuart Burrows, *L* Malcolm King, *Z* Patricia Schuman, *M* Marcel Vanaud, *C* Pierre Thau.

1987   Michael Hampe (D), Claus Viller (F), Herbert von Karajan (C), Wiener Philharmoniker, Chor der Wiener Staatsoper. Salzburg Festival. Sony SVD46383, 2000. *DG* Samuel Ramey, *DA* Anna Tomowa-Sintow, *DE* Julia Varady, *DO* John Aler, *L* Ferruccio Furlanetto, *Z* Kathleen Battle, *M* Alexander Malta, *C* Paata Burchuladze.

1987   Göran Järvefelt (D), Thomas Olfsson (F) Arnold Östman (C), Drottningholm Court Theatre Orchestra, Drottningholm Court Theatre Chorus. Drottningholm Court Theater, Stockholm. Philips 4400705191, 1991. *DG* Håkan Hagegård, *DA* Arleen Auger, *DE* Gwyneth Jones, *DO* Nico van der Meel, *L* Gilles Cachemaille, *Z* Barbara Bonney, *M* Bryn Terfel, *C* Kristinn Sigmundsson.

1987   Giorgio Strehler (D), Angelo Pacchetti (F), Riccardo Muti (C), Orchestra del Teatro alla Scala, Coro della Scala. Teatro della Scala, Milan. Opus Arte LS3001D, 2004. *DG* Thomas Allen, *DA* Edita Gruberová, *DE* Ann Murray, *DO* Francisco Araiza, *L* Claudio Desderi, *Z* Susanne Mentzer, *M* Natale de Carolis, *C* Sergej Koptchak.

1988   Lofti Mansouri (D), Norman Campbell (F), Peter Maag (C), Canada Opera Company Orchestra, Canada Opera Company Chorus. O'Keefe Center, Toronto. Unreleased broadcast. *DG* Gino Quilico, *DA* Carol Vaness, *DE* Rachel Yakar, *DO* Randall Outland, *L* Louis Quilico, *Z* Donna Brown, *M* Stephen Dupont, *C* Stephen Dupont.

1988   Marco Arturo Marelli (D), Hugo Käch (F), Bruno Weil (C), Orchester der Wiener Volksoper, Chor der Wiener Volksoper. Volksoper, Vienna. Unreleased broadcast. German. *DG* Bo Skovhus, *DA* Luba

Orgonasova, *DE* Eliane Coelho, *DO* Jorma Silvasti, *L* Gilles Cache-maille, *Z* Ulrike Steinsky, *M* Sebastian Holecek, *C* Kurt Rydl.

1988   Peter Wood (D), John Vernon (F), Colin Davis (C), Orchestra of the Royal Opera House, Royal Opera Chorus. Royal Opera House, London. Unreleased broadcast. *DG* Thomas Allen, *DA* Makvala Kasrash-vili, *DE* Kiri Te Kanawa, *DO* Stuart Burrows, *L* Stafford Dean, *Z* Joan Rodgers, *M* Gordon Sandison, *C* Gwynne Howell.

1989   Mario Corradi (D), Antoni Ros-Marbà (C), Orquesta Nacional de España, Coro del Teatro de la Zarzuela de Madrid. Teatro de la Zarzuela, Madrid. Unreleased broadcast. *DG* Wolfgang Brendel, *DA* Faye Robinson, *DE* Anna María González, *DO* Dalmacio González, *L* Francesco Ellero d'Artegna, *Z* Carmen González, *M* Alfonso Echeverria, *C* Jaakko Ryhänen.

1989   Mario Corradi (D), Pierpaolo Pessini (F), Hans Graf (C), Orchestra del Teatro Regio di Parma, Coro del Teatro Regio di Parma. Teatro Regio, Parma. Bel Canto Society BCS 0695, 1998. *DG* Renato Bruson, *DA* Winifred Faix-Brown, *DE* Daniela Dessì, *DO* Francisco Araiza, *L* Domenico Trimarchi, *Z* Adelina Scarabelli, *M* Marcello Crisman, *C* Matthias Hölle.

1990   Luc Bondy (D), Wolfgang Lesowsky (F), Claudio Abbado (C), Wiener Philharmoniker, Chor der Wiener Staatsoper. Theater an der Wien. Unreleased broadcast. *DG* Ruggero Raimondi, *DA* Cheryl Studer, *DE* Karita Mattila, *DO* Hans-Peter Blochwitz, *L* Lucio Gallo, *Z* Marie McLaughlin, *M* Carlos Chausson, *C* Anatolij Kotcherga (Anatoli Kotscherga).

1990   Luca Ronconi (D), Tonino del Colle (F), Riccardo Chailly (C), Orchestra del Teatro Comunale di Bologna, Coro del Teatro Comunale di Bologna. Teatro Communale, Bologna. Unreleased broadcast. *DG* Ruggero Raimondi, *DA* Jane Eaglen, *DE* Daniela Dessì, *DO* Rockwell Blake, *L* Alessandro Corbelli, *Z* Adelina Scarabelli, *M* Giovanni Furlanetto, *C* Andrea Silvestrelli.

1990   Peter Sellars (D), Werner Vogel (F), Craig Smith (C), Wiener Symphoniker, Arnold Schönberg Chor. Pepsico Summerfare Festival, SUNY Purchase, New York (filmed at ORF, Vienna). Decca B0004241–09, 1991. *DG* Eugene Perry, *DA* Dominique Labelle, *DE* Lorraine Hunt, *DO* Carroll Freeman, *L* Herbert Perry, *Z* Ai Lan Zhu, *M* Elmore James, *C* James Patterson.

1990   Franco Zeffirelli (D), Brian Large (F), James Levine (C), Metropolitan Opera Orchestra, Metropolitan Opera Chorus. Metropolitan Opera

House, New York. *Met Opera on Demand*, www.metopera.org/season/on-demand. *DG* Samuel Ramey, *DA* Carol Vaness, *DE* Karita Mattila, *DO* Jerry Hadley, *L* Ferruccio Furlanetto, *Z* Dawn Upshaw, *M* Philip Cokorinos, *C* Kurt Moll.

1991 Michael Hampe (D), José Montes-Baquer (F), James Conlon (C), Gürzenich Symphoniker, Chor der Oper Köln. Cologne City Opera. Arthaus 100021, 2000. *DG* Thomas Allen, *DA* Carolyn James, *DE* Carol Vaness, *DO* Kjell Magnus Sandvé, *L* Ferruccio Furlanetto, *Z* Andrea Rost, *M* Reinhard Dorn, *C* Matthias Hölle.

1991 Lindy Hume (D), Peter Butler (F), György Fischer (C), Australian Opera and Ballet Orchestra, Australian Opera Chorus. Sydney Opera House. Kultur D4307, 2009. *DG* Jeffrey Black, *DA* Gillian Sullivan, *DE* Lisa Gasteen, *DO* David Hobson, *L* Stephen Bennet, *Z* Fiona Janes, *M* Douglas McNicol, *C* Donald Shanks.

1991 Matthias Langhoff (D), Michel Dami (F), Armin Jordan (C), Orchestre de la Suisse Romande, Chœur du Grand Théâtre de Genève. Grand Théâtre de Genève. Unreleased broadcast. *DG* Thomas Hampson, *DA* Marilyn Mims, *DE* Nancy Gustafson, *DO* Gregory Kunde, *L* Willard White, *Z* Della Jones, *M* François Harismendy, *C* Carsten Harboe Stabell.

1991 David Radok (D), Michal Caban (F), Charles Mackerras (C), Orchestra of the Prague National Theater, Chorus of the Prague National Theater. Estates Theater, Prague. Supraphon SU 7012–9, 2006. *DG* Andrei Beschasny, *DA* Nadezdha Petrenko, *DE* Jiřina Marková, *DO* Vladimír Doležal, *L* Luděk Vele, *Z* Alice Randová, *M* Zdeněk Harvánek, *C* Dalibor Jedlička.

1992 Johannes Schaaf (D), Bernard Haitink (C), Orchestra of the Royal Opera House, Royal Opera Chorus. Royal Opera House, London. Unreleased broadcast. *DG* Thomas Allen, *DA* Carol Vaness, *DE* Patricia Schuman, *DO* Hans Peter Blochwitz, *L* Claudio Desderi, *Z* Marta Marquez, *M* Bryn Terfel, *C* Robert Lloyd.

1994 John Eliot Gardiner (D and C), Pim Marks (F), English Baroque Soloists, Monteverdi Choir. Concertgebouw, Amsterdam (Semi-Staged Production). Unreleased broadcast. *DG* Rodney Gilfry, *DA* Luba Orgonosova, *DE* Charlotte Margiano, *DO* Christoph Prégardien, *L* Ildebrando D'Arcangelo, *Z* Eirian James, *M* Julian Clarkson, *C* Andrew Silvestrelli.

1995 Deborah Warner (D), Derek Bailey (F), Yakov Kreizberg (C), Orchestra of the Age of Enlightenment, Glyndebourne Festival Chorus.

Glyndebourne Festival. Kultur D2032, 2003. *DG* Gilles Cachemaille, *DA* Hillevi Martinpelto, *DE* Adrianne Pieczonka, *DO* John Mark Ainsley, *L* Steven Page, *Z* Juliane Banse, *M* Roberto Scaltriti, *C* Gudjon Oskarsson.

1997 Lorenzo Mariani (D), Manuela Crivelli (F), Claudio Abbado (C), Chamber Orchestra of Europe, Coro di Ferrara Musica. Teatro Comunale di Ferrara. Unreleased broadcast. *DG* Simon Keenlyside, *DA* Carmela Remigio, *DE* Anna Caterina Antonacci, *DO* Bruno Lazzaretti, *L* Bryn Terfel, *Z* Patrizia Pace, *M* Ildebrando D'Arcangelo, *C* Matti Salminen.

1998 Achim Freyer (D), Claus Viller (F), Thomas Hengelbrock (C), Balthasar-Neumann-Ensemble, Kammerchor der Musikhochschule Heidelberg-Mannheim. Schwetzingen Festival. Unreleased broadcast. *DG* Michael Volle, *DA* Julia Isaev, *DE* Barbara Haveman, *DO* Kobie van Rensburg, *L* Renato Girolami, *Z* Sophie Koch, *M* Evert Sooster, *C* Daniel Borowski.

1999 Roberto de Simone (D), Brian Large (F), Riccardo Muti (C), Orchester der Wiener Staatsoper, Chor der Wiener Staatsoper. Vienna State Opera. TDK DVWW-OPDG, 2005. *DG* Carlos Álvarez, *DA* Adrianne Pieczonka, *DE* Anna Caterina Antonacci, *DO* Michael Schade, *L* Ildebrando D'Arcangelo, *Z* Angelika Kirchschlager, *M* Lorenzo Regazzo, *C* Franz-Joseph Selig.

2000 Franco Zeffirelli (D), Gary Halvorsen (F), James Levine (C), Metropolitan Opera Orchestra, Metropolitan Opera Chorus. Metropolitan Opera House. Deutsche Grammophon B0004051–09, 2005. Also available via *Met Opera on Demand*, www.metopera.org/season/on-demand. *DG* Bryn Terfel, *DA* Renée Fleming, *DE* Solveig Kringelborn, *DO* Paul Groves, *L* Ferruccio Furlanetto, *Z* Hei-Kyung Hong, *M* John Relyea, *C* Sergej Koptchak.

2001 Enrico Castiglione (D, F), Michael Halász (C), Orchestra Filarmonica di Roma, Coro Lirico Sinfonico Romano. Teatro Argentina, Rome. Kultur D2840, 2006. *DG* Renato Bruson, *DA* Anna Laura Longo, *DE* Amarilli Nizza, *DO* Luca Canonici, *L* Stefano de Peppo, *Z* Rossana Potenza, *M* Alessandro Battiato, *C* Nikolay Bikov.

2001 Jürgen Flimm (D), Brian Large (F), Nikolaus Harnoncourt (C), Orchester der Oper Zürich, Chor der Oper Zürich. Opernhaus Zürich. Arthaus 100329, 2002. *DG* Rodney Gilfry, *DA* Isabel Rey, *DE* Cecilia Bartoli, *DO* Roberto Saccà, *L* László Polgár, *Z* Liliana Nikiteanu, *M* Oliver Widmer, *C* Matti Salminen.

2002 Calixto Bieito (D), Toni Bargalló (F), Bertrand de Billy (C), Orchestra Academy of the Gran Teatre del Liceu, Chamber Choir of the Palau de la Música Catalana. Gran Teatre del Liceu, Barcelona. Opus Arte OA0921D, 2006. *DG* Wojtek Drabowicz, *DA* Regina Schörg, *DE* Véronique Gens, *DO* Marcel Reijans, *L* Kwangchul Youn, *Z* Marisa Martins, *M* Felipe Bou, *C* Anatoly Kocherga (Anatoli Kotscherga).

2002 Peter Brook (D), Vincent Bataillon (F), Daniel Harding (C), Mahler Chamber Orchestra, Chœur de l'Académie Européenne de Musique. Aix-en-Provence Festival. BelAir Classiques BAC210, 2006. *DG* Peter Mattei, *DA* Alexandra Deshorties, *DE* Mireille Delunsch, *DO* Mark Padmore, *L* Gilles Cachemaille, *Z* Lisa Larsson, *M* Nathan Berg, *C* Gudjon Oskarsson.

2005 Lluis Pasqual (D), Robin Lough (F), Victor Pablo Pérez (C), Orchestra of the Teatro Real, Chorus of the Teatro Real. Teatro Real, Madrid. Opus Arte OA 0958D, 2006. *DG* Carlos Álvarez, *DA* María Bayo, *DE* Sonia Ganassi, *DO* José Bros, *L* Lorenzo Regazzo, *Z* María José Moreno, *M* José Antonio López, *C* Alfred Reiter.

2006 Sven-Eric Bechtolf (D), Felix Breisach (F), Franz Welser-Möst (C), Orchester der Oper Zürich, Chor der Oper Zürich. Opernhaus Zürich. EMI Classics 50999 50097098, 2007. *DG* Simon Keenlyside, *DA* Eva Mei, *DE* Malin Hartelius, *DO* Piotr Beczala, *L* Anton Scharinger, *Z* Martina Janková, *M* Reinhard Mayr, *C* Alfred Muff.

2006 Vincent Boussard (D), Georg Wübboldt (F), René Jacobs (C), Freiburger Barockorchester, Innsbruck Festival Chorus. Innsbruck Festival. Harmonia Mundi HMD9909013, 2007. *DG* Johannes Weisser, *DA* Malin Byström, *DE* Alexandrina Pendatchanska, *DO* Werner Güra, *L* Marcos Fink, *Z* Sunhae Im, *M* Nikolay Borchev, *C* Alessandro Guerzoni.

2006 Martin Kušej (D), Karina Fibich (F), Daniel Harding (C), Wiener Philharmoniker, Konzertvereinigung Wiener Staatsopern Chor. Salzburg Festival. Decca 0743162, 2006. *DG* Thomas Hampson, *DA* Christine Schäfer, *DE* Melanie Diener, *DO* Piotr Beczala, *L* Ildebrando D'Arcangelo, *Z* Isabel Bayrakdarian, *M* Luca Pisaroni, *C* Robert Lloyd.

2007 Jossi Wieler (D) and Sergio Morabito (D), Misjel Vermeiren (F), Ingo Metzmacher (C), Nederlands Kamerorkest, Nederlands Operakoor. Netherlands Opera. Kultur D0701, 2008. *DG* Pietro Spagnoli, *DA* Myrtò Papatanasiu, *DE* Charlotte Margiano, *DO* Marcel Reijans,

L José Fardilha, *Z* Cora Burggraaf, *M* Roberto Accurso, *C* Mario Luperi.

2008   Claus Guth (D), Brian Large (F), Bertrand de Billy (C), Wiener Philharmoniker, Konzertvereinigung Wiener Staatsopern Chor. Salzburg Festival. Euroarts 2072548, 2010. *DG* Christopher Maltman, *DA* Annette Dasch, *DE* Dorothea Röschmann, *DO* Matthew Polenzani, *L* Erwin Schrott, *Z* Ekaterina Siurina, *M* Alex Esposito, *C* Anatoli Kotscherga.

2008   Francesca Zambello (D), Robin Lough (F), Charles Mackerras (C), Orchestra of the Royal Opera House, Royal Opera Chorus. Royal Opera House. Opus Arte OA 1009D, 2009. *DG* Simon Keenlyside, *DA* Marina Poplavskaya, *DE* Joyce DiDonato, *DO* Ramón Vargas, *L* Kyle Ketelsen, *Z* Miah Persson, *M* Robert Gleadow, *C* Eric Halfvarson.

2009   Pier Luigi Pizzi (D), Davide Mancini (F), Riccardo Frizza (C), Fondazione Orchestra Regionale delle Marche, Coro Lirico Marchigiano "Vincenzo Bellini." Sferisterio Opera Festival. Unitel Classica/C Major Entertainment 717504, 2014. *DG* Ildebrando D'Arcangelo, *DA* Myrtò Papatanasiu, *DE* Carmela Remigio, *DO* Marlin Miller, *L* Andrea Concetti, *Z* Manuela Bisceglie, *M* William Corrò, *C* Enrico Iori.

2010   Jonathan Kent (D), Peter Maniura (F), Vladimir Jurowski (C), Orchestra of the Age of Enlightenment, Glyndebourne Chorus. Glyndebourne Festival. EMI Classics 50999 0 72017 9 0, 2011. *DG* Gerald Finley, *DA* Anna Samuil, *DE* Kate Royal, *DO* William Burden, *L* Luca Pisaroni, *Z* Anna Virovlansky, *M* Guido Loconsolo, *C* Brindley Sherratt.

2010   Dmitri Tcherniakov (D), Andy Sommer (F), Louis Langrée (C), Freiburger Barockorchester, English Voices. Aix-en-Provence Festival. BelAir Classiques BAC080, 2011. *DG* Bo Skovhus, *DA* Marlis Petersen, *DE* Kristine Opolais, *DO* Colin Balzer, *L* Kyle Ketelsen, *Z* Kerstin Avemo, *M* David Bizic, *C* Anatoli Kotscherga.

2011   Robert Carsen (D), Patrizia Carmine (F), Daniel Barenboim (C), Orchestra del Teatra alla Scala, Coro del Teatro alla Scala. Teatro alla Scala, Milan. Deutsche Grammophon 00440 073 5221, 2015. *DG* Peter Mattei, *DA* Anna Netrebko, *DE* Barbara Frittoli, *DO* Giuseppe Filianoti, *L* Bryn Terfel, *Z* Anna Prohaska, *M* Štefan Kocán, *C* Kwangchul Youn.

2011   Michael Grandage (D), Barbara Willis Sweete (F), Fabio Luisi (C), Metropolitan Opera Orchestra, Metropolitan Opera Chorus. Metropol-

itan Opera House. *Met Opera on Demand*, www.metopera.org/season/
on-demand. *DG* Mariusz Kwiecien, *DA* Marina Rebeka, *DE* Barbara
Frittoli, *DO* Ramón Vargas, *L* Luca Pisaroni, *Z* Mojca Erdmann,
M Joshua Bloom, *C* Štefan Kocán.

2011 Göran Järvefelt (D), Cameron Kirkpatrick (F), Mark Wigglesworth
(C), Australian Opera and Ballet Orchestra, Opera Australian Chorus.
Sydney Opera House. Opera Australia POPZ56023DVD, 2012. *DG*
Teddy Tahu Rhodes, *DA* Rachelle Durkin, *DE* Jacqueline Dark,
DO Henry Choo, *L* Conal Coad, *Z* Taryn Fiebig, *M* Andrew Jones,
C Daniel Sumegi.

2011 Thilo Reinhardt (D), Volker Grohskopf (F), Johannes Wildner (C),
Orchesterakademie auf Schloss Esterházy, Opernfestspiele St. Marga-
rethen. Steinbruch St. Margarethen. Gramol 003565–1, 2011. *DG* Josef
Wagner, *DA* Mandy Fredrich, *DE* Cristina Baggio, *DO* Benjamin
Bruns, *L* Thorsten Grümbel, *Z* Anna Virovlansky, *M* Selcuk Cara,
C Dirk Aleschus.

2013 Philipp Himmelmann (D), Jérémie Cuvillier (F), Thomas Hengel-
brock (C), Balthasar-Neumann-Ensemble, Balthasar-Neumann-Chor.
Festspielhaus, Baden-Baden. Sony Classical 8843040119, 2014. *DG*
Erwin Schrott, *DA* Anna Netrebko, *DE* Malena Ernman, *DO* Charles
Castronovo, *L* Luca Pisaroni, *Z* Katija Dragojevic, *M* Jonathan Le-
malu, *C* Mario Luperi.

2014 Sven-Eric Bechtolf (D), Andreas Morell (F), Christoph Eschenbach
(C), Wiener Philharmoniker, Philharmonia Chor Wien. Salzburg Festival. EuroArts 2072738, 2015. *DG* Ildebrando D'Arcangelo, *DA* Lenneke Ruiten, *DE* Anett Frisch, *DO* Andrew Staples, *L* Luca Pisaroni,
Z Valentina Naforniţa, *M* Alessio Arduini, *C* Tomasz Konieczny.

2014 Kasper Holten (D), Jonathan Haswell (F), Nicola Luisotti (C), Orchestra of the Royal Opera House, Royal Opera Chorus. Royal Opera
House, London. Opus Arte 1145, 2014. *DG* Mariusz Kwiecien, *DA* Malin Byström, *DE* Véronique Gens, *DO* Antonio Poli, *L* Alex Esposito,
Z Elizabeth Watts, *M* Dawid Kimberg, *C* Alexander Tsymbalyuk.

2016 Michael Grandage (D), Matthew Diamond (F), Fabio Luisi (C), Metropolitan Opera Orchestra, Metropolitan Opera Chorus. Metropolitan
Opera House. *Met Opera on Demand*, www.metopera.org/season/
on-demand. *DG* Simon Keenlyside, *DA* Hibla Gerzmava, *DE* Malin
Byström, *DO* Paul Appleby, *L* Adam Plachetka *Z* Serena Malfi, *M* Matthew Rose, *C* Kwangchul Youn.

2017   Jiří Nekvasil (D), Brian Large (F), Plácido Domingo (C), National
Theatre Orchestra, Choir of National Theatre. Estates Theatre,
Prague. C Major Entertainment 74504, 2018. *DG* Simone Alberghini,
DA Irina Lungu, *DE* Kateřina Kněžíková, *DO* Dmitry Korchak,
L Adrian Sampetrean, *Z* Julia Novikova, *M* Jiří Bruckler, *C* Jan Štáva.

*

Bibliography

Abbate, Carolyn. *In Search of Opera*. Princeton, NJ: Princeton University Press, 2001.

————. "Music—Drastic or Gnostic?" *Critical Inquiry* 30, no. 3 (2004): 505–36.

————. "Opera; or, The Envoicing of Women." In *Musicology and Difference: Gender and Sexuality in Musical Scholarship*, edited by Ruth Solie, 225–58. Berkeley and Los Angeles: University of California Press, 1993.

Abert, Hermann. *W. A. Mozart*. Translated by Stewart Spencer. Edited by Cliff Eisen. New Haven, CT: Yale University Press, 2007. Originally published as *W. A. Mozart* (Leipzig: Breitkopf & Härtel, 1923–24).

Adelmann, Carl. *Donna Elvira (Don Juan) als Kunstideal und in ihrer Verkörperung auf der Münchener Hofbühne*. Munich: Theodor Ackermann, 1888.

Adorno, Theodor. "Opera and the Long-Playing Record." Translated by Thomas Y. Levin. In Adorno, *Essays on Music*, edited by Richard Leppert, 283–87. Berkeley and Los Angeles: University of California Press, 2002.

Allanbrook, Wye Jamison. *Rhythmic Gesture in Mozart: "Le nozze di Figaro" & "Don Giovanni."* Chicago: University of Chicago Press, 1983.

————. "Zerlina's 'Batti, Batti': A Case Study?" In *Siren Songs: Representations of Gender and Sexuality in Opera*, edited by Mary Ann Smart, 62–66. Princeton, NJ: Princeton University Press, 2000.

Allen, Thomas. *Foreign Parts: A Singer's Journal*. Revised and expanded ed. London: Methuen, 1995.

Altman, Rick. "Television/Sound." In *Studies in Entertainment: Critical Approaches to Mass Culture*, edited by Tania Modleski, 39–54. Bloomington: Indiana University Press, 1986.

André, Naomi. "From Otello to Porgy: Blackness, Masculinity, and Morality in Opera." In *Blackness in Opera*, edited by Naomi André, Karen M. Bryan, and Eric Saylor, 11–31. Urbana, IL: University of Illinois Press, 2012.

Andrews, Jean. "Don Giovanni: The Transgressor Transgressed." In *Selected Inter-*

disciplinary Essays on the Representation of the Don Juan Archetype in Myth and Culture, edited by Andrew Ginger, John Hobbs, and Huw Lewis, 93–127. Lewiston, NY: E. Mellen, 2001.

Ardoin, John. *The Furtwängler Record*. Portland, OR: Amadeus, 1994.

Armstrong, William. "The Costuming of a Prima Donna." *The Theatre Magazine*, May 1911, 152–56. Continued in *The Theatre Magazine Advertiser*, May 1911, viii.

Aspinall, Michael. "Adelina Patti—Speeds and Keys." *Record Collector* 37, no. 4 (1992): 268–73.

Auslander, Philip. *Liveness: Performance in a Mediatized Culture*. New York: Routledge, 1999.

Baker, Evan. "Alfred Roller's Production of Mozart's 'Don Giovanni': A Break in the Scenic Traditions of the Vienna Court Opera." Ph.D. diss., New York University, 1993.

Barraud, Henry. *Les cinq grands opéras*. Paris: Éditions du Seuil, 1972.

Bauer, Robert. *The New Catalogue of Historical Records, 1898–1908/09*. London: Sidgwick & Jackson, 1947.

Beardsley, Roger. "Speeds and Pitching of 78rpm Gramophone Records." London: Centre for the History and Analysis of Recorded Music, n.d. https://charm.rhul.ac.uk/history/p20_4_3.html.

Belton, John. "The Bionic Eye: Zoom Esthetics." *Cinéaste* 11, no. 1 (1980–81): 20–27.

Berlioz, Hector. *Memoirs of Hector Berlioz from 1803 to 1865, Comprising His Travels in Germany, Italy, Russia, and England*. Translated by Rachel (Scott Russell) Holmes and Eleanor Holmes. Annotated and revised by Ernest Newman. New York: A. A. Knopf, 1932.

Beynon, John. *Masculinities and Culture*. Buckingham, UK: Open University Press, 2002.

Bishop, Henry. *The Libertine: An Opera, in Two Acts, Founded on the Story of Don Juan; The Musick from Mozart's Celebrated Opera of "Don Giovanni," Adapted to the English Stage by Mr. Bishop*. London: John Miller, 1817.

Bitter, Christof. *Wandlungen in den Inszenierungsformen des Don Giovanni von 1787 bis 1928: Zur Problematik des musikalischen Theaters in Deutschland*. Regensburg: Bosse, 1961.

Bitter, Carl Hermann. *Mozart's "Don Juan" und Gluck's "Iphigenia in Tauris": Ein Versuch neuer Uebersetzungen*. Berlin: F. Schneider, 1866.

Blier, Steven. "The Fit Lady Sings." *Opera News*, July 2001, 12–17.

Blyth, Alan. *Opera on CD: The Essential Guide to the Best CD Recordings of 100 Operas*. London: Kyle Cathie, 1993.

Breidenstein, Helmut. *Mozart's Tempo-System: A Handbook for Practice and Theory*. Translated by Lionel Friend. Baden-Baden: Tectum, 2019.

Brock-Nannestad, George. "The Development of Recording Technologies." In Cook, Clarke, Leech-Wilkinson, and Rink, *The Cambridge Companion to Recorded Music*, 149–76.

Brody, Richard. "The Revelations of *Don Giovanni* at the Met." *New Yorker*, February 17,

2015. http://www.newyorker.com/culture/richard-brody/revelations-don-giovanni -met.

Brophy, Brigid. *Mozart the Dramatist: The Value of His Operas to Him, to His Age and to Us.* Rev. ed. London: Libris, 1988.

Brown, Clive. *Classical and Romantic Performing Practice, 1750–1900.* Oxford and New York: Oxford University Press, 1999.

Brownmiller, Susan. *Against Our Will: Men, Women and Rape.* New York: Simon & Schuster, 1975.

Brown-Montesano, Kristi. *Understanding the Women of Mozart's Operas.* Berkeley and Los Angeles: University of California Press, 2007.

Bulthaupt, Heinrich. *Dramaturgie der Oper.* 2nd ed. 2 vols. Leipzig: Breitkopf & Härtel, 1902.

Busoni, Ferruccio. *Von der Einheit der Musik.* Edited by Martina Weindel. Wilhelmshaven: Florian Noetzel, 2006.

Butt, John. *Playing with History: The Historical Approach to Musical Performance.* Cambridge: Cambridge University Press, 2002.

Catani, Ilio, ed. *Don Giovanni in Video: Roma, 28–31. Ottobre 1991.* Rome: Istituto di recerca per il teatro musicale, 1992.

Cather, Willa Sibert. "Three American Singers: Louise Homer, Geraldine Farrar, Olive Fremstadt." *McClure's Magazine,* December 1913, 33–48.

Caughie, John. *Television Drama: Realism, Modernism, and British Culture.* Oxford: Oxford University Press, 2000.

Challis, Bennett. "The Technique of Operatic Acting." *The Musical Quarterly* 13, no. 4 (1927): 630–45.

C.H.A.R.M. Discography. AHRC Research Centre for the History and Analysis of Recorded Music. https://charm.kcl.ac.uk/discography/disco.html.

Chuilon, Jacques. *Mattia Battistini: King of Baritones and Baritone of Kings.* Translated by E. Thomas Glasow. Lanham, MD: Scarecrow Press, 2009.

Citron, Marcia. "The Elusive Voice: Absence and Presence in Jean-Pierre Ponnelle's Film *Le nozze di Figaro.*" In *Between Opera and Cinema,* edited by Jeongwon Joe and Rose Theresa, 133–53. New York: Routledge, 2002.

———. *Opera on Screen.* New Haven, CT: Yale University Press, 2000.

———. "Subjectivity in the Opera Films of Jean-Pierre Ponnelle." *Journal of Musicology* 22, no. 2 (2005): 203–40.

Clarke, Kevin. "Like a Virgin?" *klassik.com,* November 18, 2006. http://magazin .klassik.com/konzerte/reviews.cfm?TASK=review&PID=1092.

Clément, Catherine. *Opera, or the Undoing of Women.* Translated by Betsy Wing. Minneapolis: University of Minnesota Press, 1988. Originally published as *L'opéra ou la défaite des femmes* (Paris: B. Grasset, 1979).

Cohen, Hermann. *Die dramatische Idee in Mozarts Operntexten.* Berlin: Bruno Cassirer, 1915.

Cone, John Frederick. *Adelina Patti: Queen of Hearts.* Chronology by Thomas G. Kaufman. Discography by William R. Moran. Portland, OR: Amadeus Press, 1993.

Cook, Nicholas. *Beyond the Score: Music as Performance*. Oxford and New York: Oxford University Press, 2013.

Cook, Nicholas, Eric Clarke, Daniel Leech-Wilkinson, and John Rink, eds. *The Cambridge Companion to Recorded Music*. Cambridge: Cambridge University Press, 2009.

Cowgill, Rachel. "Re-gendering the Libertine; Or, The Taming of the Rake: Lucy Vestris as Don Giovanni on the Early Nineteenth-Century London Stage." *Cambridge Opera Journal* 10, no. 1 (1998): 45–66.

Crutchfield, Will. "The Prosodic Appoggiatura in the Music of Mozart and His Contemporaries." *Journal of the American Musicological Society* 42, no. 2 (1989): 229–74.

Csobádi, Peter, and Gernot Gruber, eds. *Europäische Mythen der Neuzeit: Faust und Don Juan; Gesammelte Vorträge des Salzburger Symposions 1992*. 2 vols. Salzburg: Müller-Speiser, 1993.

Csobádi, Peter, Gernot Gruber, and Jürgen Kühnel, eds. *Das Musiktheater in den audiovisuellen Medien: ". . . ersichtlich gewordene Taten der Musik"; Vorträge und Gespräche des Salzburger Symposions 1999*. Salzburg: Müller-Speiser, 2001.

Csobádi, Peter, Gernot Gruber, Jürgen Kühnel, Ulrich Müller, and Oswald Panagl, eds. *Das Phänomen Mozart im 20. Jahrhundert: Wirkung, Verarbeitung und Vermarktung in Literatur, bildender Kunst und in den Medien; gesammelte Vorträge des Salzburger Symposions 1990*. Salzburg: Müller-Speiser, 1991.

Culshaw, John. *Ring Resounding*. New York: Viking, 1967.

Currie, James. "Postmodern Mozart and the Politics of the Mirror." In *Mozart Studies*, edited by Simon P. Keefe, 214–42. Cambridge: Cambridge University Press, 2006.

Curtis, Liane. "The Sexual Politics of Teaching Mozart's *Don Giovanni*." *NWSA Journal* 12, no. 1 (2000): 119–42.

Darnton, Robert. "Don Juanism from Below." In Miller, *Don Giovanni: Myths*, 20–35.

Day, Timothy. *A Century of Recorded Music: Listening to Musical History*. New Haven, CT: Yale University Press, 2000.

Dent, Edward. *Mozart's Operas: A Critical Study*. New York: McBride, Nast, 1913.

———. *Mozart's Operas: A Critical Study*. 2nd ed. London: Oxford University Press, 1947.

Deutsch, Otto Erich. *Mozart: A Documentary Biography*. Translated by Eric Blom, Peter Branscombe, and Jeremy Noble. Stanford, CA: Stanford University Press, 1965.

Dickie, Simon. *Cruelty and Laughter: Forgotten Comic Literature and the Unsentimental Eighteenth Century*. Chicago: University of Chicago Press, 2011.

Discography of American Historical Recordings (DAHR). American Discography Project. https://adp.library.ucsb.edu/index.php.

Du Mont, Mary. *The Mozart–Da Ponte Operas: An Annotated Bibliography*. Westport, CT: Greenwood Press, 2000.

Edge, Dexter. "Mozart's Viennese Copyists." Ph.D. diss., University of Southern California, 2001.

Einstein, Alfred. *Mozart: His Character, His Work*. Translated by Arthur Mendel and Nathan Broder. New York: Oxford University Press, 1945.

Eisenberg, Evan. *The Recording Angel: Music, Records and Culture from Aristotle to Zappa*. 2nd ed. New Haven, CT: Yale University Press, 2005.

Ellis, John. *Visible Fictions: Cinema, Television, Video.* Rev. ed. New York: Routledge, 1992.

Ellis, Katharine. "Rewriting *Don Giovanni*, or 'The Thieving Magpies.'" *Journal of the Royal Musical Association* 119, no. 2 (1994): 212–50.

Elste, Martin. "Vom Mythos der Realität: *Don Giovanni* als soap opera? Über Peter Sellars' Fernsehproduktion." In Csobádi and Gruber, *Europäische Mythen der Neuzeit,* 1:335–47.

Esse, Melina. "Don't Look Now: Opera, Liveness, and the Televisual." *Opera Quarterly* 26, no. 1 (2010): 81–95.

Everist, Mark. "Enshrining Mozart: *Don Giovanni* and the Viardot Circle." *19th-Century Music* 25, nos. 2–3 (2001–2): 165–89.

Ferguson, Frances. "Rape and the Rise of the Novel." *Representations* 20 (1987): 88–112.

Finck, Abbie H. C. "Maurice Renaud." *The Century,* February 1909, 611–14.

Finck, Henry Theophilus. "The Evolution of Sex in Mind." *Independent,* December 26, 1901, 3059–64.

———. *Success in Music and How It Is Won.* New York: Charles Scribner's Sons, 1909.

Fink, Gottfried Wilhelm. "Ueber das Bedürfniss, Mozarts Hauptwerke unserer Zeit so metronomisirt zu liefern, wie der Meister selbst sie ausführen leiss." *Allgemeine musikalische Zeitung* 41, no. 25 (June 19, 1839): 477–81.

Fink, Robert. "'Rigoroso ($\flat$ = 126)': *The Rite of Spring* and the Forging of a Modernist Performing Style." *Journal of the American Musicological Society* 52, no. 2 (1999): 299–362.

Fleeger, Jennifer. *Mismatched Women: The Siren's Song through the Machine.* Oxford: Oxford University Press, 2014.

Ford, Charles. *Così? Sexual Politics in Mozart's Operas.* Manchester: Manchester University Press, 1991.

———. *Music, Sexuality and the Enlightenment in Mozart's "Figaro," "Don Giovanni" and "Così fan tutte."* Burlington, VT: Ashgate, 2012.

Forney, Kristine, Andrew Dell'Antonio, and Joseph Machlis. *The Enjoyment of Music.* 12th shorter ed. New York: W. W. Norton, 2015.

Freitas, Roger. "Singing Herself: Adelina Patti and the Performance of Femininity." *Journal of the American Musicological Society* 71, no. 2 (2018): 287–369.

Gaisberg, Frederick William. *The Music Goes Round.* New York: Arno Press, [1942] 1977.

Gelatt, Roland. *The Fabulous Phonograph, 1877–1977.* 2nd ed. New York: Macmillan, 1977.

Gerstenberg, Walter. "Authentische Tempi für Mozarts 'Don Giovanni'?" *Mozart-Jahrbuch* (1960–61): 58–61.

Gielen, Michael. "Über die Tempi in Mozarts *Don Giovanni*." In program booklet, *Don Giovanni,* Oper Frankfurt, Frankfurt, September 17, 1977, 16–20.

Glass, Herbert. "'Don Giovanni'—as Perplexing as Ever." *Los Angeles Times,* September 29, 1991, Calendar section.

Goldovsky, Boris. *Bringing Opera to Life: Operatic Acting and Stage Direction.* New York: Appleton-Century-Crofts, 1968.

Gounod, Charles. *Mozart's "Don Giovanni": A Commentary*. Translated by Windeyer Clark and J. R. Hutchinson. London: R. Cocks, 1895.

Grandaur, Franz. *Don Juan: Nach dem italienischen Text des Lorenzo da Ponte für die deutsche Bühne neu bearbeitet*. Munich: Theodor Ackermann, 1874.

Gronda, Giovanna, ed. *Il Don Giovanni: Dramma giocoso in due atti; Poesia di Lorenzo Da Ponte; Musica di Mozart; Edizione critica di Giovanna Gronda*. Torino: Einaudi, 1995.

Grubb, Sufi Raj. *Music Makers on Record*. London: H. Hamilton, 1986.

Gruber, Gernot. *Mozart and Posterity*. Translated by R. S. Furness. London: Quartet, 1991.

Gruber, Paul, ed. *The Metropolitan Opera Guide to Recorded Opera*. New York: Metropolitan Opera Guild, W. W. Norton, 1993.

Guthmann, Johannes. *Goldene Frucht: Begegnungen mit Menschen, Gärten und Häusern*. Tübingen: R. Wunderlich, 1955.

Hampe, Michael. *Alles Theater: Reden und Aufsätze*. Cologne: Wienand, 2000.

Hayem, Armand. *Le Don Juanisme*. Paris: Alphonse Lemerre, 1886.

Harnoncourt, Nikolaus. *The Musical Dialogue: Thoughts on Monteverdi, Bach and Mozart*. Translated by Mary O'Neill. Portland, OR: Amadeus, 1989.

Hartford, Kassandra L. "Beyond the Trigger Warning: Teaching Operas That Depict Sexual Violence." *Journal of Music History Pedagogy* 7, no. 1 (2016): 19–34.

Helfgot, Daniel, with William O. Beeman. *The Third Line: The Opera Performer as Interpreter*. New York: Schirmer, 1993.

Henson, Karen. "Verdi, Victor Maurel and *fin-de-siècle* Operatic Performance." *Cambridge Opera Journal* 19, no. 1 (2007): 59–84.

Herz, Joachim. "Von der Bühne auf den Bildschirm: Die grundsätzlichen Möglichkeiten, die praktischen Unmöglichkeiten." In Csobádi, Gruber, and Kühnel, *Das Musiktheater in den audiovisuellen Medien*, 152–62.

Heuß, Alfred. "Das dämonische Element in Mozart's Werken." *Zeitschrift der internationalen Musikgesellschaft* 7, no. 5 (1905–6): 175–86.

Heyer, Paul. "Live from the Met: Digital Broadcast Cinema, Medium Theory, and Opera for the Masses." *Canadian Journal of Communication* 33, no. 4 (2008): 591–604.

Hicks, Alan. *Singer and Actor: Acting Technique and the Operatic Performer*. Milwaukee, WI: Amadeus, 2011.

Higgins, John. *The Making of an Opera: "Don Giovanni" at Glyndebourne*. New York: Atheneum, 1978.

Hoffmann, Ernst Theodor Amadeus. "Don Juan: Eine fabelhafte Begebenheit, die sich mit einem reisenden Enthusiasten zugetragen." In *Hoffmanns Werke in drei Bänden*, 6th ed., 1:16–29. Berlin: Aufbau-Verlag, 1979. Originally published in *Allgemeine musikalische Zeitung*, March 31, 1813.

Hogarth, Will. "A Critical Survey of the Recorded Legacy of Antonio Scotti." *Record Collector* 28, nos. 9–10 (1984): 206–27.

Hotho, Heinrich Gustav. *Vorstudien für Leben und Kunst*. Stuttgart: J. G. Cotta, 1835.

Hunter, Mary. *Mozart's Operas: A Companion*. New Haven, CT: Yale University Press, 2008.

Hurst, P. G. "Collector's Corner." *Gramophone*, February 1935, 374.

Imiela, Hans-Jürgen, and Berthold Roland. *Slevogt und Mozart: Werke von M. Slevogt zu den Opern "Don Giovanni" und "Die Zauberflöte."* Mainz: P. von Zabern, 1991.

Jahn, Otto. *W. A. Mozart.* 4 vols. Leipzig: Breitkopf & Härtel, 1856–59.

Jouve, Pierre Jean. *Le Don Juan de Mozart.* Paris: W. Egloff, 1942.

Jung-Kaiser, Ute. "'Der weiße d'Andrade': Max Slevogts verbindliche Antwort auf die Frage, wie Mozarts *Don Giovanni* zu interpretieren sei." In Csobádi, Gruber, Kühnel, Müller, and Panagl, *Das Phänomen Mozart*, 381–99.

Kalbeck, Max. *Mozart's Don Juan: Nach dem italienischen Original des Da Ponte für die deutsche Bühne frei bearbeitet und mit einem Vorwort versehen.* Vienna: A. J. Gutmann, 1886.

———. *Opern-Abende: Beiträge zur Geschichte und Kritik der Oper.* 2 vols. Berlin: Harmonie, 1898.

Katz, Mark. *Capturing Sound: How Technology Has Changed Music.* Rev. ed. Berkeley and Los Angeles: University of California Press, 2010.

Kauffmann, Stanley. "The Abduction from the Theater: Mozart Opera on Film." In *On Mozart*, edited by James M. Morris, 227–39. Cambridge: Cambridge University Press, 1994.

Keiler, Allan. *Marian Anderson: A Singer's Journey.* Champaign-Urbana: University of Illinois Press, 2002.

Kelly, Thomas Forrest. *Music Then and Now.* New York: W. W. Norton, 2013.

Kenney, William Howland. *Recorded Music in American Life: The Phonograph and Popular Memory, 1890–1945.* Oxford and New York: Oxford University Press, 2003.

Kerman, Joseph. "Reading *Don Giovanni.*" In Miller, *Don Giovanni: Myths*, 108–25.

Kierkegaard, Søren. *Either/Or.* 2 vols. Translated and edited by Howard V. Hong and Edna H. Hong. Princeton, NJ: Princeton University Press, 1987.

Kimmel, Michael. *Manhood in America: A Cultural History.* 3rd ed. Oxford and New York: Oxford University Press, 2012.

Kittler, Friedrich. *Gramophone, Film, Typewriter.* Translated by Geoffrey Winthrop-Young and Michael Wutz. Stanford, CA: Stanford University Press, 1999.

Klein, Herman. *Herman Klein and the Gramophone: Being a Series of Essays on "The Bel Canto" (1923), "The Gramophone and the Singer" (1924–1934) and Reviews of New Classical Vocal Recordings (1925–1934), and Other Writings from "The Gramophone."* Edited by William R. Moran. Portland, OR: Amadeus, 1990.

———. "The Supremacy of Mozart." *Gramophone*, December 1925, 300–02.

Klinger, Barbara. "In Retrospect: Film Studies Today." *Yale Journal of Criticism* 2, no. 1 (1988): 129–51.

Koestenbaum, Wayne. *The Queen's Throat: Opera, Homosexuality, and the Mystery of Desire.* New York: Poseidon Press, 1993.

Kristeva, Julia. *Tales of Love.* Translated by Leon S. Roudiez. New York: Columbia University Press, 1987.

Kühnel, Jürgen. "Oper im Fernsehen." In Nuy and Lemke, *Theaterbühne, Fernsehbilder*, 159–88.

———. "'Mimesis' und 'Diegesis'—Szenische Darstellung und filmische Erzählung: Zur Ästhetik der Oper in Film und Fernsehen." In Csobádi, Gruber, and Kühnel, *Das Musiktheater in den audiovisuellen Medien*, 60–79.

Kunze, Stefan. *Mozarts Opern*. Stuttgart: Reclam, 1996.

Lacey, Kate. *Listening Publics: The Politics and Experience of Listening in the Media Age*. Cambridge: Polity, 2013.

La Grange, Henry-Louis de. *Gustav Mahler*. Vol. 3, *Vienna: Triumph and Disillusion (1904–1907)*. Oxford and New York: Oxford University Press, 1999.

Lapsley, Robert, and Michael Westlake. *Film Theory: An Introduction*. 2nd ed. Manchester: Manchester University Press, 2006.

Lears, T. J. Jackson. *No Place of Grace: Antimodernism and the Transformation of American Culture, 1880–1920*. Chicago: University of Chicago Press, 1994.

Leech-Wilkinson, Daniel. *The Changing Sound of Music: Approaches to Studying Recorded Musical Performance*. London: Centre for the History and Analysis of Recorded Music, 2009. http://www.charm.rhul.ac.uk/studies/chapters/intro.html.

Lehmann, Hans-Thies. *Postdramatic Theatre*. Translated by Karen Jürs-Munby. Abingdon: Routledge, 2006. Originally published as *Postdramatisches Theater* (Frankfurt am Main: Verlag der Autoren, 1999).

Lehmann, Lilli. *Mein Weg*. 2nd ed. Leipzig: G. Hirzel, 1920.

Leinsdorf, Erich. *The Composer's Advocate: A Radical Orthodoxy for Musicians*. New Haven, CT: Yale University Press, 1981.

Leppert, Richard. *Aesthetic Technologies of Modernity, Subjectivity, and Nature: Opera, Orchestra, Phonograph, Film*. Berkeley and Los Angeles: University of California Press, 2015.

Lert, Ernst. *Mozart auf dem Theater*. Berlin: Schuster & Loeffler, 1918.

Levi, Erik. *Mozart and the Nazis: How the Third Reich Abused a Cultural Icon*. New Haven, CT: Yale University Press, 2010.

Levi, Hermann. *Mozarts Don Juan: Der bestrafte Wüstling oder Don Juan; Heiteres Drama in zwei Akten; Dichtung von Lorenzo Da Ponte; Der italienische Originaltext, mit hauptsächlicher Benützung der Übersetzung von Franz Grandaur, neu bearbeitet von Hermann Levi*. Munich: Theodor Ackermann, 1897.

Levin, David. *Unsettling Opera: Staging Mozart, Verdi, Wagner, and Zemlinsky*. Chicago: University of Chicago Press, 2007.

Lipking, Lawrence. "Donna Abbandonata." In Miller, *Don Giovanni: Myths*, 36–47.

Littlejohn, David. *The Ultimate Art: Essays Around and About Opera*. Berkeley and Los Angeles: University of California Press, 1992.

Lustig, Larry. "Charles Gilibert." *Record Collector* 64, no. 1 (2019): 46–55.

Malloch, William. "Toward a 'New' (Old) Minuet." *Opus*, August 1985, 14–21, 52.

Mann, Thomas. *The Magic Mountain*. Translated by H. T. Lowe-Porter. New York: A. A. Knopf, 1927.

Mann, William. *The Operas of Mozart*. Oxford and New York: Oxford University Press, 1977.

Marty, Jean-Pierre. *The Tempo Indications of Mozart*. New Haven, CT: Yale University Press, 1988.

Maurel, Victor. *Dix ans de carrière, 1887–1897*. Paris: P. Dupont, 1897.

McCracken, Allison. *Real Men Don't Sing: Crooning in American Culture*. Durham and London: Duke University Press, 2015.

Meyer, Stephen C. "*Parsifal*'s Aura." *19th-Century Music* 33, no. 2 (2009): 151–72.

———. "Sound Recording and the End of the Italian *Lohengrin*." *Cambridge Opera Journal* 20, no. 1 (2008): 1–24.

Mila, Massimo. *Lettura del Don Giovanni di Mozart*. Turin: Einaudi, 1988.

Miller, Jonathan, ed. *Don Giovanni: Myths of Seduction and Betrayal*. Baltimore, MD: Johns Hopkins University Press, 1990.

Miller, Karl Hagstrom. *Segregating Sound: Inventing Folk and Pop Music in the Age of Jim Crow*. Durham, NC: Duke University Press, 2010.

Milner, Greg. *Perfecting Sound Forever: An Aural History of Recorded Music*. New York: Farrar, Straus & Giroux, 2009.

Moberly, Robert Basil. *Three Mozart Operas: "Figaro," "Don Giovanni, "The Magic Flute."* New York: Dodd, Mead, 1968.

Morley, David. *Television, Audiences and Cultural Studies*. London: Routledge, 1992.

Morris, Christopher. "Digital Diva: Opera on Video." *Opera Quarterly* 26, no. 1 (2010): 96–119.

Mozart, Leopold. *A Treatise on the Fundamental Principles of Violin Playing*. Translated by Editha Knocker. 2nd ed. Oxford and New York: Oxford University Press, 1985.

Mozart, Wolfgang Amadeus. *Don Giovanni*. Edited by Wolfgang Plath and Wolfgang Rehm. *Neue Mozart Ausgabe* ser. II, cat. 5, vol. 17. Kassel: Bärenreiter, 1968. *Kritischer Bericht*, presented by Wolfgang Rehm, 2003.

———. *Don Giovanni*. Edited by Alfred Einstein. London: Eulenburg, n.d.

———. *Don Giovanni*. Edited by Georg Schünemann and Kurt Soldan. Leipzig: C. F. Peters, 1941. Reprint, New York: Dover, 1974.

———. *Don Giovanni, An Opera, in Two Acts: Translated into English and Corrected According to the Original Score by Natalia MacFarren*. London: Novello, n.d.

———. *Don Giovanni, ossia, Il dissoluto punito: Dramma giocoso in due atti, posto in musica e ridotto per il cembalo*. Paris: Maurice Schlesinger, 1822.

———. *Don Juan: Musique de Mozart; Grand Opéra Français en 5 actes; Paroles Françaises de MM. Émile Deschamps et H. Blaze*. Paris: V. Launer, n.d.

———. *Il dissoluto punito ossia il Don Giovanni: Facsimile of the Autograph Score*. Introductory Essay by Hans Joachim Kreutzer. Musicological Introduction by Wolfgang Rehm. 3 vols. Los Altos, CA: Packard Humanities Institute, 2009.

Mulvey, Laura. "Visual Pleasure and Narrative Cinema." *Screen* 16, no. 3 (1975): 6–18.

Neumann, Frederick. "The Appoggiatura in Mozart's Recitative." *Journal of the American Musicological Society* 35, no. 1 (1982): 115–37.

———. "Improper Appoggiaturas in the *Neue Mozart Ausgabe*." *Journal of Musicology* 10, no. 4 (1992): 505–21.

Nuy, Sandra, and Inga Lemke, eds. *Theaterbühne, Fernsehbilder: Sprech-, Musik- und Tanztheater im und für das Fernsehen*. Salzburg: Müller-Speiser, 1998.

Osborne, Charles. *The Complete Operas of Mozart: A Critical Guide*. New York: Atheneum, 1978. Reprint, New York: Da Capo Press, 1983.

Oulibicheff, Alexandre [Ulybyshev, Aleksandr Dmitrievich]. *Nouvelle biographie de Mozart, suivie d'un aperçu sur l'histoire générale de la musique et de l'analyse des principales oeuvres de Mozart*. 3 vols. Moscow: Auguste Semen, 1842–43.

Ovid. *The Erotic Poems*. Translated by Peter Green. London: Penguin, 1982.

Parakilas, James. "The Afterlife of *Don Giovanni*: Turning Production History into Criticism." *Journal of Musicology* 8, no. 2 (1990): 251–65.

Parker, Dorothy. "The Artist's Reward." *New Yorker*, November 30, 1929, 28–29.

Parker, Roger. *Remaking the Song: Operatic Visions and Revisions from Handel to Berio*. Berkeley and Los Angeles: University of California Press, 2006.

Pfeiffer, K. Ludwig. "'Directed for Video': Zum Stellenwert der Inszenierungen Peter Sellars." In Csobádi, Gruber, and Kühnel, *Das Musiktheater in den audiovisuellen Medien*, 172–83.

Phelan, Nancy. *Charles Mackerras: A Musician's Musician*. London: V. Gollancz, 1987.

Philip, Robert. *Early Recordings and Musical Style: Changing Tastes in Instrumental Performance, 1900–1950*. Cambridge: Cambridge University Press, 1992.

———. *Performing Music in the Age of Recording*. New Haven, CT: Yale University Press, 2004.

Pinza, Ezio. *An Autobiography*. Written with Robert Magidoff. New York: Rinehart, 1958.

Poriss, Hilary. *Changing the Score: Arias, Prima Donnas, and the Authority of Performance*. Oxford and New York: Oxford University Press, 2009.

Possart, Ernst. *Ueber die Neueinstudierung und Neuszenierung des Mozart'schen Don Giovanni (Don Juan) auf dem kgl. Residenztheater zu München*. Munich: A. Bruckmann, 1896.

Rank, Otto. *The Don Juan Legend*. Edited and translated by David G. Winter. Princeton, NJ: Princeton University Press, 1975. Originally published as *Die Don-Juan Gestalt* (Leipzig: Internationaler Psychoanalytischer Verlag, 1924).

Rice, John A. *Mozart on the Stage*. Cambridge: Cambridge University Press, 2009.

Rochlitz, Friedrich. *Don Juan: Oper in Zwei Akten; Nach dem italienischen des Abb. da Ponte frei bearbeitet*. Leipzig: Breitkopf & Härtel, 1801.

Rockwell, John. "A 'Giovanni' Reborn as 'Donna.'" *New York Times*, July 18, 1987.

Rohringer, Stefan. "Don Ottavio and the History of the Tenor Voice." In Julian Rushton, Stefan Rohringer, Sergio Durante, and James Webster, *Dramma Giocoso: Four Contemporary Perspectives on the Mozart/Da Ponte Operas*, edited by Darla Crispin, 33–58. Leuven: Leuven University Press, 2012.

Romain, Michael. *A Profile of Jonathan Miller*. Cambridge: Cambridge University Press, 1992.

Rose, Brian G. *Television and the Performing Arts: A Handbook and Reference Guide to American Cultural Programming*. Westport, CT: Greenwood, 1986.

Rosenberg, Alfons. *Don Giovanni: Mozarts Oper und Don Juans Gestalt*. Munich: Prestel, 1968.

Ross, Alex. "The Storm of Style: Listening to the Complete Mozart." *New Yorker*, July 24, 2006. http://www.newyorker.com/magazine/2006/07/24/the-storm-of-style.

Roujon, Henry. *Miremonde*. Preface by Alexandre Dumas *fils*. Paris: Ollendorff, 1896.

Rous, Samuel Holland. *The Victor Book of the Opera: Stories of One-Hundred Operas with Five-Hundred Illustrations & Descriptions of One-Thousand Victor Opera Records*. Revised ed. Camden, NJ: Victor Talking Machine Company, 1913.

Rudolf, Max. "Ein Beitrag zur Geschichte der Temponahme bei Mozart." *Mozart-Jahrbuch* (1976–77): 204–24.

Rushton, Julian. *W. A. Mozart: "Don Giovanni."* Cambridge Opera Handbooks. Cambridge: Cambridge University Press, 1981.

Rutherford, Susan. *The Prima Donna and Opera, 1815–1930*. Cambridge: Cambridge University Press, 2006.

Schmidt Horning, Susan. *Chasing Sound: Technology, Culture, and the Art of Studio Recording from Edison to the LP*. Baltimore, MD: Johns Hopkins University Press, 2013.

Schmidt-Sistermanns, Johannes. *Opernregie im Fernsehen: Medienspezifische Regiekonzepte zur Visualisierung von Oper im Fernsehen*. Vienna: Verband der Wissenschaftlichen Gesellschaften Österreichs, 1991.

Schönhaar, Rainer. "Mozart als Filmkomponist? Am Beispiel *Don Giovanni* 1954/1979." In Csobádi, Gruber, Kühnel, Müller, and Panagl, *Das Phänomen Mozart*, 407–32.

Schor, Naomi. *Reading in Detail: Aesthetics and the Feminine*. New York: Methuen, 1987.

Schwarzkopf, Elisabeth, J. B. Steane, and Alan Sanders. *Elisabeth Schwarzkopf: A Career on Record*. Portland, OR: Amadeus, 1995.

Segal, Lynne. *Slow Motion: Changing Masculinities, Changing Men*. 3rd ed. Basingstoke, UK: Palgrave Macmillan, 2006.

Senici, Emanuele. "Porn Style? Space and Time in Live Opera Videos." *Opera Quarterly* 26, no. 1 (2010): 63–80.

Shaw, Bernard. *Shaw's Music: The Complete Musical Criticism*. Edited by Dan H. Laurence. 3 vols. London: M. Reinhardt and The Bodley Head, 1981.

Shea, George Edward. *Acting in Opera: Its A–B–C; With Descriptive Examples, Practical Hints and Numerous Illustrations*. New York: G. Schirmer, 1915.

Sheppard, W. Anthony. "Review of the Metropolitan Opera's New HD Movie Theater Broadcasts." *American Music* 25, no. 3 (2007): 383–87.

Shuker, Roy. *Wax Trash and Vinyl Treasures: Record Collecting as Social Practice*. Burlington, VT: Ashgate, 2010.

Smart, Mary Ann. "Defrosting Instructions: A Response." *Cambridge Opera Journal* 16, no. 3 (2004): 311–18.

Steane, J. B. *The Grand Tradition: Seventy Years of Singing on Record*. 2nd ed. Portland, OR: Amadeus, 1993.

Steiger, Karsten. *Opern-Diskographie: Verzeichnis aller Audio- und Video-Gesamtaufnahmen*. 2nd expanded ed. Munich: K. G. Saur, 2008.

Stiftung Mozarteum Salzburg. *Mozart-Bibliographie*. https://mozarteum.at/en/bibliotheca-mozartiana/#mozart-bibliography-section.

Stivender, David, and Susan Webb, translators. *"Don Giovanni"* [English translation]. In *The Metropolitan Opera Book of Mozart Operas*, edited by Paul Gruber, 297–381. New York: HarperCollins, 1991.

Swarowsky, Hans. *Wahrung der Gestalt: Schriften über Werk und Wiedergabe, Stil und Interpretation in der Musik*. Edited by Manfred Huss. Vienna: Universal Edition, 1979.

Sussman, Herbert. *Masculine Identities: The History and Meanings of Manliness*. Santa Barbara, CA: Praeger, 2012.

Sutcliffe, Tom. *Believing in Opera*. Princeton, NJ: Princeton University Press, 1996.

Tambling, Jeremy. *Opera, Ideology and Film*. New York: St. Martin's Press, 1987.

————, ed. *A Night in at the Opera: Media Representations of Opera*. London: J. Libbey, 1994.

Taruskin, Richard. *Text and Act: Essays on Music and Performance*. Oxford: Oxford University Press, 1995.

Thoma, Hubertus. "Maria Cebotari." Translated by David Mason. *Record Collector* 64, no. 4 (2019): 226–57.

Tirso de Molina. *The Trickster of Seville and the Stone Guest (El burlador de Sevilla y el convivado de piedra)*. Translated by Gwynne Edwards. Reprint ed. Oxford: Oxbow Books, 2016.

Tommasini, Anthony. "Second Date with a Little Black Dress." *New York Times*, June 11, 2008. https://www.nytimes.com/2008/06/11/arts/music/11voig.html.

————. "Should the Fat Lady Diet Before She Sings?" *New York Times*, November 14, 2004. https://www.nytimes.com/2004/11/14/arts/music/should-the-fat-lady-diet-before-she-sings.html.

Tosh, John. *Manliness and Masculinities in Nineteenth-Century Britain: Essays on Gender, Family and Empire*. Harlow: Pearson Longman, 2005.

Trezise, Simon. "The Recorded Document: Interpretation and Discography." In Cook, Clarke, Leech-Wilkinson, and Rink, *The Cambridge Companion to Recorded Music*, 186–208.

Türk, Daniel. *Klavierschule, oder Anweisung zum Klavierspielen für Lehrer und Lerndende, mit kritischen Anmerkungen*. Leipzig and Halle: Schwickert, 1789.

Turner, Kristen M. "Class, Race, and Uplift in the Opera House: Theodore Drury and His Company Cross the Color Line." *Journal of Musicological Research* 34, no. 4 (2015): 320–51.

Villadsen, Lisa Storm. "The Rhetoric of Seduction." Ph.D. diss., Northwestern University, 2000.

Wallnig, Josef. "'Die Bewegung unter allen Umständen genau treffen, erfordert lange Übung' (D. G. Türk): Überlegungen zur Temponahme von Mozarts *Don Giovanni*." *Österreichische Musikzeitschrift* 45, no. 5 (1990): 227–33.

Walter, Bruno. *Theme and Variations: An Autobiography*. Translated by James A. Galston. New York: Alfred A. Knopf, 1946.

Welch, Walter L., and Leah Brodbeck Stenzel Burt. *From Tinfoil to Stereo: The Acoustic Years of the Recording Industry, 1877–1929*. Gainesville: University Press of Florida, 1994.

Werckmeister, Johanna. "Die zweite Regie: Formen audiovisueller Adaption des Musiktheaters. Dargestellt am Beispiel von Bizets *Carmen*." In Nuy and Lemke, *Theaterbühne, Fernsehbilder*, 189–206.

Wiesmann, Sigrid, and Ulrich Müller. "Gespräch mit Brian Large über Probleme der Oper im Fernsehen." In Csobádi, Gruber, and Kühnel, *Das Musiktheater in den audiovisuellen Medien*, 44–59.

Wilentz, Sean. *360 Sound: The Columbia Records Story*. San Francisco: Chronicle Books, 2012.

Will, Richard. "The Ambivalence of Mozart's Countess." In *Music, Libraries, Scholar-

ship: Essays in Honor of Lenore Coral, edited by James P. Cassaro, 31–54. Middleton, WI: A-R Editions, 2007.

———. "*Don Giovanni* and the Resilience of Rape Culture." *Journal of the American Musicological Society* 71, no. 1 (2018): 218–22.

———. "Zooming In, Gazing Back: *Don Giovanni* on TV." *Opera Quarterly* 27, no. 1 (2011): 32–65.

Williams, Bernard. "Don Giovanni as an Idea." In Williams, *On Opera*, 31-42. New Haven, CT: Yale University Press, 2006. Originally published in Rushton, *W. A. Mozart: "Don Giovanni,"* 81–91.

Wlaschin, Ken. *Encyclopedia of Opera on Screen: A Guide to More than 100 Years of Opera Films, Videos, and DVDs*. New Haven, CT: Yale University Press, 2004.

Wolzogen, Alfred Freiherrn von. *Wilhelmine Schröder-Devrient: Ein Beitrag zur Geschichte des musikalischen Dramas*. Leipzig: F. A. Brockhaus, 1863.

Woodfield, Ian. *Performing Operas for Mozart: Impresarios, Singers and Troupes*. Cambridge: Cambridge University Press, 2012.

———. *The Vienna "Don Giovanni."* Woodbridge: Boydell Press, 2010.

Worthen, William B. *Shakespeare Performance Studies*. Cambridge: Cambridge University Press, 2014.

Wright, Sarah. "Consuming Passions: The Aesthetics of Cultural Consumption in Calixto Bieito's *Don Giovanni* (2001)." *Journal of Spanish Cultural Studies* 5, no. 3 (2004): 317–39.

Zaslaw, Neal. Mozart's Symphonies: Context, Performance Practice, Reception. Oxford: Clarendon Press, 1989.

Index